SPARKS OF LIFE

SPARKS OF LIFE

Darwinism and the Victorian Debates over Spontaneous Generation

JAMES E. STRICK

HARVARD UNIVERSITY PRESS
Cambridge, Massachusetts
London, England
2000

Printed in the United States of America

Library of Congress Cataloging-in-Publication Data

Strick, James Edgar, 1956–
Sparks of life : Darwinism and the Victorian debates over spontaneous generation / James E. Strick.
p. cm.
Includes bibliographical references (p.) and index.
ISBN 0-674-00292-X (cloth : alk. paper)
1. Spontaneous generation—Great Britain—History—Victoria, 1837–1901. 2. Evolution—Great Britain—History—Victoria, 1837–1901. I. Title.

QH325 .S85 2000
576.8′8—dc21 00-031915

To Dr. Chester M. Raphael, whose life
and work have been an inspiring example.
With gratitude for many years of encouragement
and advice, and with deep respect for his integrity,
this work is dedicated to him.

ACKNOWLEDGMENTS

This book was first conceived as a doctoral dissertation in the Program in History of Science at Princeton University. For an enormous amount of advice and guidance in that project, I am grateful to Dr. Gerald Geison, a true scholar in this field. It is to Gerry that I owe my first insight into thinking historically about spontaneous generation, as well as the initial encouragement to pursue training in this field. The close reading and editorial suggestions of Dr. Lynn Nyhart have also been of enormous help. Both of them have helped to make the book much more readable and the argument much clearer.

A book like this requires the assistance of many dedicated people and institutions. In the Princeton History of Science Program the help of Peggy Reilly and Kathy Baima was always offered generously. At the Wellcome Institute Bill Bynum and Sally Bragg were very generous with office space, resources, and advice. Archivists in many institutions were enormously helpful and friendly. I would like to mention particularly Anne Barrett (the Huxley papers), Frank James and Irene McCabe (the Tyndall papers), Cheryl Piggott at Kew, Richard Aspin at the Wellcome Institute London, Gina Douglas and Sarah Darwin at the Linnean Society Library, Steve Wilson at the National Hospital Queen Square, Eileen Massey at the American Philosophical Society, and Malcolm Beasley at the British Museum of Natural History. My thanks to all these archives for permission to quote from unpublished manuscript materials. I also wish to thank Academic Press and Kluwer Academic Publishers for permission to reprint some of this material that appeared in a different form in the *Encyclopedia of Microbiology*, 2nd ed. and the *Journal of the History of Biology*, respectively.

For helpful discussions and feedback throughout the research, writing, and rewriting, I am grateful to many friends and scholars. Particularly perceptive and helpful criticism came from Ruth Barton, Bill

Bynum, Michael Collie, Adrian Desmond, Bernie Lightman, and Jim Moore. But I would also like to thank Gar Allen, Toby Appel, Tom Broman, Angela Creager, John Dettloff, Otni Dror, John Farley, Sophie Forgan, Dan Gilroy, Sander Gliboff, Jan Golinski, Jack Haas, Joseph Heckman, Mary Higgins, Larry Holmes, Teri Hopper, L. S. Jacyna, Frank James, Harmke Kamminga, Stuart McCook, Jane Maienschein, Andrew Mendelsohn, Stanley Miller, David Nanney, Bob Olby, Mercer Rang, Nick Rasmussen, Marsha Richmond, Nils Roll-Hansen, Terrie Romano, Jim Secord, Grier Sellers, Philip Sloan, Bill Summers, Nancy Tomes, Maria Trumpler, Sonia Uyterhoeven, Mick Worboys, Alan Yoshioka, and the Princeton History of Science Seminar. Of course, any errors that remain are my sole responsibility.

Much of the research was supported by a National Science Foundation dissertation improvement grant. Much of the writing was supported by a Dibner Institute post-doctoral fellowship and the Biology and Society Program of Arizona State University. To Jane Maienschein, the co-director of that program, I am deeply grateful for much mentoring and personal friendship. I also wish to express my gratitude to Jim Collins, chair of the biology department at Arizona State University, for having the vision to include history of science as an integral part of a functioning science department.

A project of this size also attests to the existence of extensive personal support. My friend and teacher David Brahinsky knows as well as anyone what I hoped to accomplish, and he gave selflessly of love, work, and knowledge on my behalf. Sue Bannon, Oliver LoPresti, Amy Sabsowitz, Dan Silverman, and Cara Tripodi gave much to me during the time when I wondered whether I could ever complete such an undertaking. Most important, my wife Wendy Sobey and our daughter Rachel good-naturedly suffered the loss of my time and attention that this work cost. And both of them have never stopped believing in me, through book revisions, colic episodes, or desert heat.

CONTENTS

SPARKS OF LIFE

Introduction

This book is about scientific debates from 1860 to 1880 in Britain over the origin of life by spontaneous generation. It is also about how those debates were inextricably intertwined with the public controversy over Darwin's theory of evolution in the first two decades after the publication of his *On the Origin of Species* in 1859. The idea that living things can originate suddenly from nonliving materials, spontaneous generation, has a long history that is inseparably intertwined as well with the development of microbiology and the germ theory of disease. Thus, this book is also about important steps in the emergence of microbiology as a laboratory science and of the relations between that science and medicine.

Past histories of the debates of 1860–1880 have overwhelmingly emphasized the experiments alone. While experiments are undeniably important in the story, this book will emphasize previously understudied aspects of theory change that were also important in the debates.[1] Furthermore, I will show that these disputes cannot be understood outside the context of factional in-fighting among the Darwinians themselves, as they struggled to create a socially and scientifically viable form of "Darwinian" science.

The aim of this book is to answer the following historical questions: first, how could spontaneous generation come to take root for a time (even a brief one) in so thoroughly unsuitable a soil as British natural theology? Second, why did spontaneous generation then lose favor so decisively after 1880? My argument is that several different clusters of observations and ideas were sources of spontaneous generation theories and pillars of support for them. Only when, one by one, *all* of the pillars were removed did spontaneous generation theories finally fall as far out of favor as they have been during the twentieth century.

No less an authority than Aristotle claimed that cases of spontaneous generation could be observed in nature, and his support was important in establishing the idea for many centuries. Beginning around the time of the Scientific Revolution of the seventeenth century, however, the doctrine of spontaneous generation was increasingly challenged and became the subject of numerous episodes of controversy. As combatants tried to answer one another's criticisms, the new breakthroughs in technique and in experimental design that were developed served as some of the most important foundations upon which a science of microbiology could be built. As just one example, the development of sterile technique as well as procedures to sterilize glassware and growth media all grew directly out of experiments trying to prove or disprove the possibility of spontaneous generation of microorganisms.

In the search to answer a question as basic as how life originates, theory played a role as important or more so than technique or experiment. From the first, the doctrine of spontaneous generation was seen to be fraught with religious implications. If life could originate spontaneously from lifeless matter, the position of philosophical materialism, then a Creator God was irrelevant. If spontaneous generation occurred in present times, this was also at odds with a single original Creation as described in the Bible. However, those interested in a naturalistic worldview, such as supporters of Darwinian evolution, had a potential conflict as well. The doctrine of evolution was based on a profound philosophical assumption of continuity in nature, i.e., that there were no sudden unbridgeable gaps between similar living forms, which would require supernatural intervention. Furthermore, Darwin's theory implied that the vast diversity of living things had come from one or at most a few first ancestral organisms, and these must have originated somehow. For many, then, to believe in evolution and a completely naturalistic worldview required the belief that no unbridgeable gap occurred between living matter and nonliving matter and that living organisms must have been capable of arising from nonlife at least once on the early earth. Henry Charlton Bastian (the last great supporter of spontaneous generation and a central figure in this story) and others with this view were evolutionists, and they charged hypocrisy on the part of those who supported Darwin but were unwilling to believe in the necessity of spontaneous generation.

Still another important source of disagreement, even among scientists who claimed not to care about religious implications, were fundamental epistemological assumptions about the nature of life. Some believed very deeply that all living things must reproduce by "seeds" or "germs," by analogy with the large number of organisms for which this process had been observed. So fundamental was this belief that, even in cases where microscopic life appeared in tubes of fluid boiled for hours, those scientists concluded that either the tubes must have leaked after boiling, or that there must be some kind of structures produced by microorganisms that were capable of withstanding previously unheard of temperatures, even though nobody had ever seen such structures. Their opponents, as late as the 1870s, believed that the most clearly observed and reliably established fact known about living things was that they were totally unable to survive the boiling temperature of water for more than a few minutes. From this equally tenable premise, they concluded that spontaneous generation was a less strained explanation of organisms in boiled infusions than the *ad hoc* invoking of "spores" (that nobody had ever experimentally demonstrated) with totally unprecedented heat-resisting abilities. Thus, before ever carrying out an experiment, both sides in the spontaneous generation controversies were often begging the question at issue. Researching these philosophical issues better demonstrates the lack of communication among experimenters than the differences in the experiments themselves. Looking at the larger context of philosophical issues also explains why the doctrine of spontaneous generation has risen and fallen in popularity repeatedly at different times in different countries in the past several centuries. To tell the story of spontaneous generation controversies only as a series of battles about "dueling experiments" would be to misunderstand much or even most of what the controversies were really about.

A look at some case histories from earlier spontaneous generation debates will illustrate this point. This brief survey will also highlight the usefulness of historical methods in clarifying even the technical points that supposedly were involved.

Early History

From antiquity it had been maintained that frogs, eels, mice, and numerous worms and insects, especially parasitic worms living inside animals, could arise by spontaneous generation. With Leeuwenhoek's discovery of microorganisms, many naturalists assumed that these too could arise without parents. Indeed, as microbes seemed exceedingly simple, and bacteria simplest of all, it was believed that they were the most likely to be organizable from nonliving materials. Francesco Redi, natural philosopher to the same Tuscan court that was patron to Galileo, carried out some famous experiments in 1668 that investigated specifically the origin of insects. Redi's experiments showing that maggots come from fly eggs, not from rotting meat, usually lead off histories of spontaneous generation debates. It was commonly believed until that time that the appearance of maggots in rotting meat was a clear example of spontaneous generation. Redi placed samples of many different types of meat and fish in glass jars. One set of jars was open to the air and soon swarmed with maggots. The other set was covered with fine muslin cloth. Redi saw that, while maggots never appeared in the meat of those jars, flies crawled about on the cloth and sometimes laid eggs there. Those eggs were seen to hatch into maggots, disproving spontaneous generation as their origin. Though Redi himself continued to believe that some insects, such as gall flies, did arise by spontaneous generation, many naturalists assumed from this time onward that spontaneous generation, if it occurred, only did so among parasitic worms and microorganisms.

Of course, working at the time of the Scientific Revolution, Redi did participate in a time when experiment came to the fore as an activity ever afterwards central in natural philosophy. Spontaneous generation debates, like all other natural philosophic questions, featured experiments in a more and more important role from the late seventeenth century onward. But just as "natural history" or "natural philosophy" historically did not mean what we now call "science" until fairly recently, neither did "experiment" always mean what it means now. Many natural philosophers in the seventeenth century were interested in the power of the experimental method, but were just as interested in public demonstrations of "experiments" as a way of convincing audiences, e.g., prospective patrons, that their enterprise was qualitatively different from the book-dominated natural philosophy of the past, and from the subjective and often bloody religious disputes that had traumatized European life for over a century. Recent studies have looked closely at Redi's career, especially his relationship to the Medici Grand Dukes, his principal patrons. One such study, by Paula Findlen, concludes that Redi's public demonstration of experiments was a form of court entertainment in which the final arbiter of the meaning and success of the outcome was the Grand Duke. Thus there was a world of difference between Redi's procedures and what by the nineteenth century would be called "controlled experimentation by a biologist."[2] Redi did set high value on repeated observation and demonstration, saying, "I do not put much faith in matters not made clear to me by experiment." Paradoxically, however, he suggested elsewhere that he experimented "in order to make myself more certain of that of which I am already most certain." For Redi, then, especially in his primary role as an early modern courtier, "experiment" meant something quite different from what we understand by that term today, particularly with regard to the role of "preconceived expectations."[3] The paradox results only if we assume in a naively ahistorical way that what Redi called "experiment" should be assumed to have the same meaning that term came to have after "natural philosophy" had been transformed into "science."

Once we gain this richer sense of what experiment meant in the historical context in which Redi lived and worked, we may begin to question the validity of an ahistorical narrative that uncritically compares his work with experiments performed by Pasteur, Tyndall, and their antagonists two centuries later, merely because of superficial similarity in "the use of controls." Perhaps there are more similarities in the meaning of "experiment" in such different historical settings, but this is an open question which can only be answered by looking at the full context in which each investigator had to operate, including such questions as the existing system of patronage upon which each depended to support work on scientific pursuits.

Needham versus Spallanzani

Eighty years after Redi, in 1748, another series of experiments by the Irish priest John Turberville Needham was published, and this time was widely believed to support the spontaneous generation of microorganisms. Needham soon collaborated with the French aristocrat Comte de Buffon. And over the next several decades the work of Buffon and Needham was opposed by many, including Charles Bonnet and Lazzaro Spallanzani.

Needham's first experiments found that, when mutton gravy was stoppered and sealed with mastic in glass tubes and heated to boiling for a time, the fluid afterwards teemed with microorganisms (protists). Buffon and Needham later reported seeing tiny particles (in the size range of bacteria) that they called "organic molecules," which came from disintegrating organic material and moved very actively. These, they said, clumped together to form the larger animalcules (protists). They also claimed that sealing and boiling proved that the molecules did not originate from "insects or eggs floating in the atmosphere." Most naturalists considered this, if true, to be a case of spontaneous generation.

Spallanzani, an Italian cleric, carried out experiments challenging these claims and first published his results in 1765. His opposition was based on two main arguments. First, he said that with his own microscope he never saw anything like the "organic molecules" and therefore charged that Buffon and Needham were only seeing what they wanted to see, because Buffon's philosophy of nature had predicted that such particles must exist. This claim was widely repeated. Second, he said that when he boiled infusions similar to Needham's, he did not find any living microbes in them, as long as he sealed the tubes by melting the glass shut in a flame before boiling and continued the boiling for at least an hour. Thus, Spallanzani claimed that the microorganisms in Needham's infusions must indeed have gotten in from the atmosphere after boiling, despite Needham's precautions.

Needham responded that boiling for as much as an hour would clearly be so severe a treatment as to deprive the air in the tubes of its power to generate or support life. Later experiments showing that indeed such treatment decreased or consumed the oxygen content in the tubes seemed to support Needham's claim. Thus the experimental outcome was underdetermined by the experiments themselves, so the controversy was able to continue through the 1780s.

A popular notion of experiment has it that "proper science" can only occur when the scientist approaches his experiment with no "preconceived ideas" about the outcome, and intends to "let the chips fall where they may." In many histories of the spontaneous generation debates, the "losers" are often accused of having been biased beforehand by their belief in a "vital force" or some similar overarching philosophy, while Redi, Spallanzani, Pasteur, and Tyndall epitomize the open-minded investigator.[4] A closely related claim has been that Spallanzani, Pasteur, and others were often able to defeat their foes by virtue of possessing better instruments, especially superior microscopes. Thus, as the story has been told since even the early nineteenth century, the Comte de Buffon and his collaborator Needham in the 1740s were portrayed as "armchair philosophers" who cooked up a doctrine of "organic molecules," a vital "plastic force," and spontaneous generation; and they had such inferior microscopes that they were able to interpret whatever fuzzy images they saw as supporting the fuzzy ideas they *wanted* the data to confirm. Victorian biologist T. H. Huxley's version established one of his most famous clichés by describing Buffon and Needham's work as an object lesson for generations of young scientists: "Such as it was, I think it [Buffon and Needham's doctrine] will appear . . . to be a most ingenious and suggestive speculation. But the great tragedy of Science—the slaying of a beautiful hypothesis by an ugly fact—which is so constantly being enacted under the eyes of philosophers, was played almost immediately, for the benefit of Buffon and Needham."[5] Huxley argued, based perhaps on Pasteur's similar claim, that Buffon and Needham had a poor microscope and their opponent Spallanzani had a much more objective outlook and understanding of the method of controlled experiment.

Though seldom emphasized, it is very important to note that spontaneous generation was seen to support two other, broader doctrines: epigenesis and materialism. Epigenesis is the doctrine that all the parts of an embryo are assembled gradually, having at first been nonexistent. Epigenesis was opposed in the seventeenth and eighteenth centuries by the doctrine of preformation, which claimed that all embryonic organs existed inside the sperm or the egg in miniature from even before fertilization and only needed to grow, unfold, and expand as gestation proceeded. This implied logically that each generation of, e.g., humanity must have been contained within the previous one, like Russian dolls, all the way back to the eggs in Eve's ovaries or the sperm of Adam. (The overwhelming majority of naturalists writing during this period were Christian.) Since Buffon and Needham's description of the origin of protist microbes suggested their organization from originally homogeneous "molecules," this seemed a blow for preformationists. Buffon indeed ridiculed preformation theory, insisting that the microscope showed no trace of these successively enclosed generations within the sperm or eggs of animals. Spallanzani eventually came down on the side of preformation and rejected Needham's theory. We would probably admit today that, since his microscope could not have shown him little preformed homunculi inside eggs or sperm cells, Spallanzani, too, must have been willing to go some way in believing in things he could not verify and was thus also guilty of "philosophizing."[6]

More than that, if Buffon and Needham's observations were correct, they showed that matter contained within itself all the properties necessary to organize into life. This was the basic tenet of philosophical materialism, a doctrine profoundly at odds with Christianity or any kind of Deism. Voltaire, among others, feared that Needham's claims would support atheism and materialism. It was thought by many that this theory implied that life could originate by a chance, random combination of substances. This was so contrary to mainstream religious beliefs, and to a natural philosophy still very much in the service of demonstrating the existence of a beneficent Creator, that it generated heated opposition. Further, beginning around 1750–1770, this led to spontaneous generation becoming strongly associated with atheism, materialism, and political radicalism.

Since this chance combination of chemicals has become such a crucial element of modern views of the origin of life, such as the famous 1953 Urey-Miller experiment, it is interesting that Buffon and Needham are not celebrated as thinkers far ahead of the religious biases of their time. Voltaire seems to have been one of those who most actively spread the opinion that Needham and Buffon were poor scientists, though this was based on seriously misreading Needham.[7]

Only recently has close study been applied to the actual technical evidence on what kinds of microscopes were being used by Spallanzani and by Needham. This work, by Philip Sloan, has shown that Buffon and Needham's results cannot be successfully explained by assuming that they worked with poor instruments, nor that their work was sloppy or biased in advance by a priori theorizing, despite the fact that these explanations are ubiquitous among previous accounts of the controversy. Indeed, a close review of the evidence on the actual experiments suggests that the experiments of Needham and Buffon were careful, high quality work with quite a bit of sophisticated hypothesis formation and testing.[8]

Many of the contemporary critics of Buffon and Needham, including Spallanzani in the 1760s, had originated this claim of a priori bias at least partly *because* they assumed the pair to have worked with a British Cuff compound microscope, common at that time. This type of device had a maximum magnification of only about 100x, and even at that low power the image it produced suffered from severe chromatic and spherical aberration. Such aberration was a problem with all compound microscopes prior to the perfection of achromatic lenses, sixty years later. By careful analysis of the original publications, however, it has been shown that the instrument used by Needham was actually a high-quality single-lens microscope of the Wilson screw-barrel design, capable of at least 400x magnification with outstanding resolution, i.e., *not* plagued by the aberrations of compound scopes. This was similar to the instrument with which Robert Brown discovered the cell nucleus in plant cells and Brownian movement in pollen particles in the 1820s and 1830s. In fact, Buffon and Needham's observations anticipate those that led to the discovery of Brownian movement. Their "organic molecules" are what Brown later called "active molecules." The simple microscope employed in the famous experiments by Spallanzani was a variant of the Lyonnet aquatic microscope. Spallanzani's instrument was incapable of the short focal length, high-resolution work permitted by the Wilson screw-barrel design, so that he could not even see the bacteria-sized "organic molecules" under dispute. Spallanzani was the technically handicapped participant, though exactly the opposite story has become universal.[9] Buffon and Needham's discovery of "organic molecules" was not the work of imagination, but an actual discovery of active molecules and of Brownian movement some eighty years before Brown. Perhaps because of their faith in their evidence, Buffon and Needham refused to back down from their original observations, despite accusations of atheism and decades of controversy. Furthermore, both sides seem to have had plenty of a priori philosophical commitments at stake, and nobody approached as explosive a subject as materialism with a completely open mind.

Worms, Molecules, and Evolution

John Farley has shown that the appearance of parasitic worms within animals was seen until perhaps the 1840s as the strongest single piece of evidence supporting spontaneous generation claims. The complex life cycles of most of these parasites involved intermediate life-cycle stages that had to live in other host animals. Thus, egg-feeding experiments would not have been able to show eggs from one animal directly producing worms in another. Only in the 1840s and 1850s, with the working out of the complex life cycles and intermediate hosts, did it finally become unambiguously clear that the worms reproduced via eggs.

Despite the denunciation of Buffon and Needham, a great many pathologists, histologists, and, later, cytologists continued to observe bacteria-sized, actively moving particles in blood, tissues, and infusions. These particles continued to be called "molecules" by life scientists. They were using the term, obviously, in a way quite different from the chemists' use of the term that developed at about the same time. Important examples include the "active molecules" of Robert Brown (1828–29), the cell-forming cytoblastema "molecules" of Theodor Schwann (1839), and the histological "molecules" of John Hughes Bennett (1840–1875). Many scientists thought that the clumping of these units was the basis for heterogenesis, and that this furthermore supported epigenesis as the correct explanation of embryonic development. Especially vocal in this belief were Lorenz Oken, Karl Burdach, and other German biologists of the *Naturphilosophie* school. (It should be emphasized that neither Brown nor Schwann themselves interpreted their "molecules" in this way, and Schwann did some significant experiments early in his career that seemed to disprove certain cases of spontaneous generation.) In Britain, both Charles Darwin and Richard Owen worked in private on the molecule theory during the 1830s, as did William Addison in the 1840s. More important, the British histologist Bennett was one of the first professors to teach a physiology course in a university and medical school. From 1840 until almost 1870, he taught his theory of "molecules" as the essential units from which cells formed, explicitly embracing heterogenesis by the late 1860s. Thanks to Bennett's position as a professor, his ideas were thus widely influential among British scientific and medical circles for over a generation.

J. B. Lamarck, the early French evolutionist, had made spontaneous generation of microorganisms an integral part of his evolution theory from its beginning in 1800.[10] Lamarckian supporters were numerous and widespread, though often associated with radical politics, as self-organization and development of matter was seen as a crucial scientific underpinning for democratic, bottom-up organizing political theories. Lamarck himself had been appointed to his professorship in Paris in 1793 by the revolutionary government, as part of its reforms of the educational system.[11] Thus his ideas were widely associated with the radical politics of France in the first decade after the revolution. Detractors saw them as examples of those strains of enlightenment thought that had led to the revolution. Evolution and spontaneous generation undermined religion, went the argument, and thus the moral and social fabric.[12] Supporters of Lamarckian ideas were often enthusiasts of radical, democratizing politics. In Anglican-dominated Britain, very few such Lamarckians managed to rise to university teaching posts; however, notable among those that did was the outspoken comparative anatomy professor Robert Grant. Though considered not quite respectable in Victorian intellectual circles, his teaching of evolution and spontaneous generation continued from the 1820s through his death in 1874, and thus gave those ideas a presence in British thought decades before *On the Origin of Species* appeared. Charles Darwin, for instance, was one of Grant's earliest students and protégés at Edinburgh University (though Darwin later felt he had to avoid any contact with the less-than-respectable Grant when he was developing his own evolution theory in secret for twenty years). Henry Charlton Bastian also studied under Grant, at University College London in the late 1850s. Grant and other Lamarckians had established in the public mind a strong link between evolutionary theories and spontaneous generation, and between radical politics and both doctrines.[13]

Note on Terminology

John Farley has emphasized how much incorrect analysis has resulted from historians lumping together under the heading "spontaneous generation" two ideas that were considered significantly different by most of the participants in the debates. These are heterogenesis and abiogenesis, and Farley's distinction has greatly clarified the story of spontaneous generation disputes. According to Farley's definition, heterogenesis is the idea that living organisms can arise without parents from organic starting materials. Abiogenesis is the idea that living things can arise without parents, from simple inorganic compounds. It is important to note, however, that even Farley's distinction between these two ideas is not always identical to the usage of the actors.[14]

It is crucial for modern students of the history of science to understand the amount of confusion that was generated by the ambiguity and vagueness of many of the terms used. I show in chapter 2 that "molecules" is one of the most prominent and hitherto overlooked examples.

In addition, such terms as "vibrio," "monads," "bacteria," "confervae," "spores," "infusoria," "microzymes," and "bioplasts" were very imprecise in 1860, partly representing the early stage of development of a science of microbial life.[15] The meaning of these terms underwent constant revision in response to differing usages during these debates, especially in pleomorphist theories that saw many of these microscopic objects as interconvertible.[16] I discuss this in chapter 5. The terms "heterogenesis," "xenogenesis," "spontaneous generation," "biogenesis," "archebiosis," "germ theory of disease," and "antiseptic surgery" showed similar ambiguity and constant revision during the period of this story. Many times terms were given altered meanings, or new ones coined, in an attempt to gain the rhetorical upper hand in the debate.[17] Thus, while a glossary is provided attempting to give definitions for most of these terms, the reader should remember that the very slipperiness of the terms and the changing definitions of many of them are crucial to how the debates turned. It is only by tracking the usage of the terms closely throughout that we can see when the ambiguity of a term, rather than actual empirical observations, is a source of disagreement as well as when deliberate changes in usage represent rhetorical strategies. As Ludwik Fleck has warned, words can become "battle cries," charged by association with much more than mere observational content.[18] "Biogenesis" and "abiogenesis," two of the most familiar terms to modern readers, prove to be among the most interesting in this light, as I will show in chapter 4.

Why Another Study of the Spontaneous Generation Debates?

It is no exaggeration to say that with the possible exception of evolution, no controversy in the history of biology has generated as much secondary and popular literature as the spontaneous generation debates of the nineteenth century. The linkage of spontaneous generation with evolution was one of the most powerful pillars of support for the former doctrine during the British debates. However, very little work has been done to situate the spontaneous generation views of important evolutionists such as Herbert Spencer, T. H. Huxley, John Tyndall, J. D. Hooker, Anton Dohrn, and E. Ray Lankester in the context of their broader efforts to make evolutionary theory respectable, and to professionalize the science of biology in the wake of Chambers's *Vestiges of the Natural History of Creation*. The fact that that enormously popular work was written by an amateur naturalist was one of the chief features that made it anathema to Huxley. The book solidified in the public mind a linkage between transmutation of species and spontaneous generation. Thus, after Huxley became a supporter of Darwin's transmutationist theory, it became his goal to destroy that linkage in order to firmly separate Darwin's respectable, professional science from the transmutation theories that had come before. He worked at many different levels to accomplish this separation: through the power of his numerous influential scientific offices, in public lectures, and as part of the design of his new elementary biology course, which he hoped would reform the training of British students, especially those destined to become biologists.

Huxley, Tyndall, and others spent most of their professional careers trying to advance this broad agenda while moving from outsiders to positions of power within the Victorian scientific establishment, even the elite Royal Society of London. The organized functioning of the X Club was central to the success of these broader goals, thus I will examine the position of the X Club toward the debates on spontaneous generation, particularly toward a previously unrecognized but significant minority faction of Darwinians in Britain who were also supporters of spontaneous generation. Huxley must have maneuvered with as much agility around "abiogenesis" as he did around "agnosticism," because of the negative associations of atheism and materialism from *opponents* of Darwinism. However, Huxley also needed to manage the rebellious faction that gathered around Bastian, as Darwinian *supporters*. This homegrown rebellion could not be attributed to the peculiarities of German thought, as has Ernst Haeckel's support for spontaneous generation, and its story as a chapter in the power struggles within British Darwinism remains to be told. This book will tell that story.

Previous literature has tended to suggest that only relatively minor figures supported spontaneous generation in Britain prior to Henry Charlton Bastian in 1870. John Farley's book-length survey is the first to introduce the role of Owen and of Bennett, and to move beyond "dueling experiments" narratives. Farley does not explain the conversion of Bennett and Owen, though, and this needs considerable analysis unless one is willing to accept a sudden collapse of the influence of natural theology in 1869 without explaining how it was weakened. I begin to reconstruct this story in chapter 2. Nicolaas Rupke does follow Owen's conversion in the late 1850s to becoming a supporter of Pouchet's heterogenesis. However, neither Rupke's biography of Owen nor Adrian Desmond's new biography of Huxley attaches much importance to Owen's support for spontaneous generation, and neither appreciates the crucial role that he had in setting the stage for the British debates from 1868 until 1880. The *Lancet* and *British Medical Journal,* not to mention some evolutionists, now found Owen taking what they considered the most cutting-edge anticlerical position, drastically altering the party lines that had solidified during the previous thirty years. The debate over spontaneous generation during the 1860s and 1870s, then, was not a minor sidelight to other battles in British science. It was a major axis around which dramatic new alignments were being defined. In-fighting occurred between evolutionary factions, between medical practitioners and newly professionalizing laboratory biologists and physiologists, and between proponents of the germ theory of disease and their opponents across many disciplinary boundaries.

To date, only Alison Adam's analysis begins to use the lens of the spontaneous generation debates to study these major shifts and realignments in British science and medicine.[19] She outlines the tension that developed between medical practitioners and non–life scientists such as Tyndall. However, she still portrays Bastian as a fluke in Britain, and fails to recognize that the "scientific naturalism" of Huxley and Tyndall and the older natural theology tradition were not the only contenders in the debates. The category of scientific naturalism as employed by Adam, for all its analytic usefulness, is too internalist to capture the crucial dimensions of the spontaneous generation story. Bastian shared almost all the basic tenets of scientific naturalism as Adam defines it. As I argue in chapter 3, Bastian was viewed as one of the most promising of the young Darwinian stable until he publicly broached the spontaneous generation question against Huxley's advice.

The activities of the X Club make a more useful tool with which to analyze the controversy, I show in chapter 4, because this allows the entire dimension of "proper scientific behavior," so crucial to Huxley's agenda of professionalized science, to assume a role in the story. More than that, it invites a study of the new science journals being introduced during this time. *Nature,* for instance, was the long-sought journal that the "Young Guard" had hoped would act as the organ for the new evolutionary science, and as a voice of cultural authority. However, within eight months of its creation, it was playing out Huxley's worst nightmare, fragmenting rebellious factions away from the X Club line, and in full public view. A whole new genre of popular science journals for middle-brow audiences was just springing up during the 1860s as well, and several of these, such as *Scientific Opinion*, also took a maverick line from the X Club over spontaneous generation. The spontaneous generation debates can shed light upon who defined the nature of science in the late nineteenth century, and which version of that science was considered authoritative.

In that sense, the medical journals, though more professional than in the 1830s, were still quite radical. As Tyndall expressed it to Darwin, anticipating whipping them into line on the issue of spontaneous generation: "The change in the medical journals is radical—they see that the end of the nonsense which they have so long countenanced is nigh." To properly evaluate the resistance of the British medical profession to Pasteur's pronouncements on spontaneous generation and on the germ theory of disease, a further look is needed into the relative weight of personality clashes, disciplinary independence, and opinions about the experiments.

Margaret Pelling's work establishes critical context in explicating the alternative to Pasteur, based on Liebig's chemistry, that most British practitioners interested in "scientific medicine" relied upon to explain zymotic disease. But in addition to theory, many physicians found the doctrine of spontaneous generation appealing, not least because it was in keeping with traditional medical practice. By contrast, disproof of spontaneous generation was thought by many to imply the victory of the germ theory, which in turn suggested the intrusion of bacteriology and experimental science in general into the clinician's traditional domain. As I demonstrate in chapter 6, not only did early simplistic versions of the germ theory seem to ignore any role for the patient's constitution, but, for Pasteur, diseases ought to have been totally redefined—by their causative germ, not by their symptoms.

All very interesting, it might be argued, but still, why invoke all these other factors when the experimental evidence alone is enough to explain the outcome of the debates? There are two answers to this question. The first is that, since all of these other disputes were in fact going on during the spontaneous generation debates, the story is much richer and more complete as cultural history if they are included as part of it. But, for those who would claim that this is still not "part of the science," I must emphasize that even the technical aspects of spontaneous generation experiments have been dealt with too simplistically. Harry Collins has written much in recent years about the "experimenter's regress," or the problem of knowing or negotiating when an experiment has really been completed. In the spontaneous generation experiments of the 1860s and 1870s, the question of heat-resistance of microorganisms and their spores was one of the most important unknowns being studied, and each investigator's theoretical assumptions obviously had a great influence on the design of the experiments and their interpretation. The previous literature on this question, e.g., by John Crellin and Glenn Vandervliet, for all of its thorough attention to the chronology of new technical developments (Cohn's discovery of spores in hay and cheese bacilli, Tyndall's fractional sterilization, and so on), has given little attention to the intensity of the negotiations that took place *before* these discoveries.[20] At that time, without any agreed-upon way to decide that sufficient precautions had been taken to sterilize infusions by heating, Pasteur and Tyndall were nonetheless sure that their opponents' infusions must not have been treated with adequate care, begging the very question at issue.

In 1974, Farley and Geison began to explore this point and to exploit it as a measure of the relative strength of social and political forces in Pasteur's science. They point out that, had Pasteur used Pouchet's hay infusions in his famous swan-necked flasks, the story might have had a very different outcome. Farley's book-length treatment shows that Tyndall made a major rhetorical blunder requiring an embarrassing public turnaround after the discovery of heat-resistant spores. The implication was clear that his earlier results had been exaggerated in the service of his preconceived belief in the impossibility of spontaneous generation. Bastian did not hesitate to point this out at the time, and questions remained about whether Tyndall had actually shown hay spores to be the cause of the growth in infusions; but Tyndall's change of front was nonetheless successful. Similarly, Bastian showed at the time that Huxley was wrong to claim Bastian did not know enough to distinguish between Brownian movement and true living movement in his infusions. (In fact Bastian had been the first to publish the importance of that distinction himself, a few months earlier.) Nonetheless, Huxley was able to make his accusation stick. Chapters 4 and 7 explore the background context of the X Club's tactics, especially at the Royal Society, in order to explain how such misrepresentations of the facts succeeded. An exactly analogous experimental debate between Bastian and Pasteur over urine infusions was never settled experimentally. As Geison shows in his biography of Pasteur, it is necessary to explore the larger context of Pasteur's connections with the French Academy of Sciences, the church, and the state, to understand why the Academy nonetheless considered Pasteur to have won. Thus, even with regard to the details of the experiments themselves, since it can be shown that there was always some degree of underdetermination, it is necessary to develop the larger context in order to properly evaluate to what extent the experimental evidence was the decisive factor.

If Tyndall was right in the end experimentally about heat-resistant spores, it meant that he had been wrong all along in his belief that Bastian could not possibly be a competent experimenter. Many former supporters of Bastian were persuaded by spores and finally accepted with regret that Bastian (while competent) had been mistaken. Yet Tyndall's own version of the story never adopted this point of view. As a result an incorrect picture of Bastian as a pitiful figure, both experimentally incompetent and too stubborn to admit it, still persists widely in the literature on this subject. Farley began the process of dismantling this portrait, and Adam carried this further. But only in the full context of Bastian's role as leader of a rebellious evolutionary faction can a still more symmetrical story take shape.[21]

In this book, I seek not only to tell the story of Bastian. His figure dominates the story of the British debates from 1870 onward, and his personality as well as Tyndall's and Huxley's became bound up inseparably with their theoretical views in the process. But that story remains subservient to the larger aim of explicating what spontaneous generation itself meant to British scientists between 1860 and 1880. I have alluded to many cultural and disciplinary developments against which spontaneous generation had meaning, but important theoretical changes in biology are also part of the picture. Numerous authors have pointed out that development, generation, inheritance, and evolutionary theory were all part of the same problem complex until early in the twentieth century.

To fully understand the implications of the spontaneous generation debates it is necessary to reconstruct an even larger problem complex, in which the emerging concepts of conservation of energy and correlation of forces were also seen as bound up with evolutionary theory. I outline some of the features of this complex in chapter 5. In addition, I show that concepts not previously seen as relevant to these debates must be brought in, including Brownian movement, colloids, and a doctrine from histology and pathology about "histological molecules" that reached its most complete expression in the "molecular theory of organization" of John Hughes Bennett. Bennett argued this doctrine was more fundamental than the cell theory as Virchow had formulated it, especially with reference to the formation of pathological cells. For many, Darwin's 1868 theory of pangenesis was seen within the discourse about histological molecules. The demise of this doctrine went along with major changes in the understanding of the ontological status of the *chemist's* molecules and atoms, Brownian movement, cell theory, the relationship of colloids to living cells, the nature of bacteria, and the role of the cell nucleus in inheritance and cell multiplication.

Each of these changes contributed to a major theoretical realignment in biology, and each in a specific way removed part of the theoretical framework in which spontaneous generation could seem possible. Without the major theoretical changes in life sciences, the underdetermined experiments alone could not have brought down spontaneous generation as completely as it fell by 1880. These realignments were crucial to resolving the "experimenter's regress" in favor of Tyndall's experiments and against those of Bennett and Bastian in the minds of a decisive majority, as were the maneuvers of the X Club in London scientific politics. The outcome of the controversy contributed to a dramatic change in the relationship between clinical medicine and the laboratory medical sciences, with physiology and bacteriology in dominant roles at much more of a distance from medical practitioners than might otherwise have been the case. That is the still-unexplored story that this book seeks to tell. I will begin, for the sake of comparison, with the standard story of the debates.

1

Spontaneous Generation and Early Victorian Science

The Standard Story of the British Debates

How living things originate, especially the earliest ones, was a question of great interest to Charles Darwin. Especially after the publication of *On the Origin of Species* in 1859, many in Darwin's audience, enthusiastic about his totally naturalistic method of explaining life, wondered whether this also implied a non-miraculous process for the origin of life. Historically, reasoning and experiment on this question had been highly contentious and usually fell under the rubric of "spontaneous generation." Advocates of spontaneous generation claimed that organisms could sometimes arise by the right combination of nonliving materials under appropriate conditions. Opponents claimed this was never possible; that living things must always come from parents like themselves. The term "spontaneous generation" was considered antiquated and simplistic by most of the principals in the debate by the late 1860s. As John Farley has pointed out, it also obscures a crucial distinction drawn by most of them: that between what was commonly called "heterogenesis" or "heterogeny" and what was called "abiogenesis" or "archebiosis." Heterogenesis is the process of living things allegedly appearing from degenerating material that itself was derived from previously living things (e.g., meat or vegetable infusions). Archebiosis is the process of living things allegedly appearing from inorganic starting materials. Many believed in the occurrence of one of these processes but not the other, for various reasons. Some journals, especially those aimed at a nonscientific public, continued to describe any supporter of either doctrine as an advocate of spontaneous generation, thus often lumping together individuals with significant disagreements over one another's views.

Because it implied the possibility of a universe without any necessity for a Creator God, spontaneous generation (especially archebiosis) was opposed strongly by many scientists functioning in the Christian cultures of Western Europe. Darwin's theory of transmutation by natural selection encountered a very similar reaction for the same reason. There was an implicit logical connection between the two theories: if the origins of all living things could be traced to fewer and fewer common ancestors, then the origins of a few or perhaps even one original organism still required a naturalistic explanation. In addition to this, Lamarck had popularized an earlier theory of transmutation in his *Philosophie Zoologique* in which present-day spontaneous generation played an integral part.

For Darwin and some of his followers such as T. H. Huxley and Herbert Spencer, present-day spontaneous generation was not necessarily required for evolution, even if they did believe that in the distant past it may have occurred.[1] Thus, when Louis Pasteur's experiments of 1860–1862 were reported, Darwin, Huxley, and Spencer considered the matter settled. Pasteur had carried out experiments that persuaded many that present-day spontaneous generation was impossible, routing his opponent Felix Pouchet and receiving a prize from the French Academy of Sciences for finally settling the question, in its judgment.[2] Pouchet had claimed to show that bacteria could appear in previously boiled and sealed infusions made from, e.g., hay. Pasteur had shown with other infusions (he did not use hay) that various ingenious precautions which prevented dust from entering his flasks and tubes could reliably prevent any growth in them. Most noteworthy among these methods were his famous "swan-necked flasks," which allowed air to enter, but trapped dust in the twists and turns of the long drawn-out glass neck.

In the powerful natural theology tradition of British science up to this time, it is a tribute to Darwin's open-mindedness that he was willing to consider the possibility of spontaneous generation. No other prominent British scientist of the time would have considered Pouchet's doctrine worth looking at, let alone worthy of support. Robert E. Grant, one of the few British Lamarckians who did support spontaneous generation, proves the point.[3] His was an unorthodox voice whose science was clearly merged with his radical political interests; indeed, Huxley said "within the ranks of the biologists at that time [1850s], I met with nobody, except Dr. Grant of University College, who had a word to say for evolution—and his advocacy was not calculated to advance the cause."[4]

That the Darwinians considered the question settled by Pasteur was clearly stated in Huxley's 1870 presidential address to the British Association for the Advancement of Science (BAAS).[5] Huxley had even undertaken some experimental work himself to follow up Pasteur's and to understand in more detail the nature of protoplasm, which in 1868 he had dubbed "the physical basis of life." He thus led British thought on the subject, stating conclusively that although abiogenesis (as he called spontaneous generation) might have occurred, indeed *probably* occurred, in the distant past, Pasteur's experiments convinced him that no such process could still be occurring today. Huxley's clear exposition was very influential, as evidenced by the use of his terms "biogenesis" and "abiogenesis" in most discussion of the subject from that point forward.

It was during this period, in 1868, that Huxley made a discovery that was considered related to the origin of life question by many. Upon reexamining deep sea sediment samples after many years, Huxley found ubiquitously in them a gelatinous substance with microscopic bodies scattered throughout. Huxley decided this was a primitive undifferentiated form of protoplasm, one of the most primordial living organisms. Because Ernst Haeckel, the German evolutionist, had recently predicted such organisms and devised the class name Monera for them, Huxley christened the creature *Bathybius haeckelii*.[6] Haeckel suggested that the organism was the product of spontaneous generation constantly occurring on the sea bottom. Many came to believe that Huxley must also accept that idea, especially when, just a few months after the discovery, he gave a famous lecture entitled "The Physical Basis of Life."

Huxley's lecture set out to describe the physical-chemical nature of protoplasm and to emphasize that all living things were composed of this same basic substance. Huxley declared that the lecture had as one goal to show the fallibility of the doctrine of materialism. He was known as a very shrewd public speaker that often cleverly undercut theology by appearing to use its own arguments, however.[7] Therefore, when his talk described protoplasm as *only* a chemical substance, one which was made by organisms from inorganic nutrients in their food, many in his audience thought he was cleverly arguing for a materialistic view of life and stopping just short of publicly affirming the implicit possibility of the spontaneous generation of protoplasm from those same inorganic chemical constituents. Even after his 1870 BAAS address, Haeckel's opinion still convinced many that Huxley really did believe modern abiogenesis possible. It was only when he announced in 1875 that *Bathybius* was a mistake, a chemical artifact of the preservation of sea sediment in alcohol, that the British public saw that Huxley's views on abiogenesis were truly limited to its occurrence in the distant past.

More typical of the British scientific tradition rooted in natural theology was William Thomson's 1871 presidential BAAS lecture. Thomson had rejected Darwinian evolution on the basis of a physicist's estimate of the earth's age, which held that age as far too young for the hundreds of millions of years required for natural selection. Now in his 1871 address, he responded to Huxley, suggesting that abiogenesis even in the distant past was unnecessary to explain the origin of life. A deeply religious man, Thomson suggested a preferable hypothesis: that germs of life could have been brought to earth from other planets, carried on meteorites!

Pasteur's work had inspired two other noted British workers, the Edinburgh surgeon Joseph Lister and the physicist and popular spokesman of science John Tyndall. Lister followed Pasteur's suggestion that the air germs that contaminate infusions must also be able to cause disease and contaminate wounds, leading to infection and putrefaction there. Lister found that spraying the antiseptic carbolic acid over a wound while performing surgery and keeping the wound covered with a carbolic acid–soaked bandage until it closed greatly reduced the chances of infection. He announced his system of antiseptic surgery in 1867, causing other doctors to begin taking Pasteur's germ theory of disease more seriously.[8] John Tyndall produced evidence supporting Pasteur's idea via a different route. While the physicist was trying to obtain dust-free air for experiments on light scattering, he was surprised to notice that the vast majority of atmospheric dust was organic in nature, since it could be removed by combustion. He announced in a famous lecture in January 1870, "Dust and Disease," that the nature of dust bore out Pasteur's and Lister's claims, and that airborne germs were probably the real source of disease, especially contagious disease.[9]

Bastian and Burdon Sanderson

Pasteur, Huxley, Tyndall, and others working on this question were, however, unaware of the existence of heat-resistant bacterial spores, found for example in the hay bacillus *Bacillus subtilis*, and in *Clostridium butyricum* from cheese. They therefore assumed that exposure to boiling temperature for a minute or two was sufficient to sterilize most infusions. Until these spores were discovered in 1876, it was possible for results in boiled infusions to mistakenly lead the unwary to believe that bacteria had been spontaneously generated there. And this is precisely what was done by Henry Charlton Bastian, a young neurologist and professor of pathological anatomy at University College London. Bastian zealously set out to prove his cherished belief in the ability of bacteria, protozoa, yeasts, and molds to be generated from nonliving materials, including even inorganic substances—what he termed "archebiosis." In his first book in 1871, Bastian attacked Huxley's 1870 position and set out his own agenda. He daringly claimed that Pasteur had wrongly interpreted his own experiments, and that the famous chemist's results were actually better explained as supporting (or at least not ruling out) the possibility of spontaneous generation.

But this was by no means the full extent of Bastian's crusade. Only a year later he published a mammoth two-volume opus of over 1,300 pages, *The Beginnings of Life*, listing countless experiments in which microbes had appeared in boiled infusions and arguing that boiling for five minutes was surely sufficient to kill any living thing. Since, again, this idea of a "thermal death point" was widely believed to be the boiling temperature of water, many read Bastian's voluminous results and wondered if, indeed, spontaneous generation might be possible. Bastian was a very lucid and persuasive writer and speaker, and he became an overnight sensation by the end of 1872. So much was this the case that one of London's most prominent scientists, John Burdon Sanderson, physiology professor at University College, decided to investigate Bastian's claims for himself.

Among the experiments Bastian reported as most successful in generating microbial life was one involving an infusion of hay and another of boiled turnip and cheese. Burdon Sanderson contacted Bastian and asked him to prepare these infusions while Burdon Sanderson observed (and in some cases even helped). The infusion was made by cooking sliced turnips in water and filtering out the particles to yield a clear broth. Then, while this broth was boiling, a few grains of pounded cheddar cheese were added. Boiling was continued for five minutes more, and the flasks were then hermetically sealed with a flame while still boiling. Burdon Sanderson requested the additional precaution, agreed to by Bastian, that the flasks should be heated at 250°C prior to filling, to insure that no live microorganisms could possibly be clinging to their inner walls and thus acting as contaminants. The contents of some flasks were naturally acidic, while others were neutralized with KOH solution ("liquor potassæ"). In all of the flasks, Burdon Sanderson reported, numerous living bacteria were to be found within a few days. Since many had doubted Bastian's experimental competence, including Burdon Sanderson himself, he wished to declare that, whatever interpretation one might choose for the facts stated by Bastian, their veracity as experimentally determined facts should not be doubted. Burdon Sanderson desired not to commit himself any further in discussing interpretation, insisting, "I have hitherto taken no part in the controversy relating to spontaneous generation, and do not intend to take any."[10] Nonetheless, Burdon Sanderson's reputation as a skilled and unbiased experimental scientist lent considerable authority to Bastian's point of view, so that several years later Charles Darwin would still say, "I do not care much about what Dr. Bastian says, but I feel *very strongly* that the whole subject is not made clear until some light is thrown on the question how men like Burdon Sanderson and Wyman of Boston and Dr. Child often succeeded in getting bacteria in infusions which they had boiled for a long time."[11]

Burdon Sanderson's generous and well-intentioned action thus combined with ignorance of heat-resistant spores to lend Bastian's point of view more credence than it deserved. Nor was Burdon Sanderson the only figure of importance make such an error. Prior to his action George Bentham, president of the Linnean Society, announced in 1872 that spontaneous generation "is still supported by so many naturalists whose opinions are entitled to consideration, and there is so much to be said for as well as against it which appears unsusceptible of direct and positive proof, that it is likely to be long maintained as a subject of controversy."[12]

Bastian himself referred to Burdon Sanderson's experiments to prove that negative results by others were not sufficient to disprove the possibility of spontaneous generation, i.e., that it is logically impossible to prove the negative: "the obtaining of such negative results is always easy, and may show nothing more than the relative incapacity of the experimenter for performing careful work according to instructions."[13]

Criticisms of Bastian by Lankester and Roberts

Numerous critics, including E. Ray Lankester and William Roberts, continued to suggest possible problems with Bastian's technique that might be the source of introducing contamination from without the flasks. Lankester suggested that Bastian was not grinding up the cheese into fine enough particles, and argued that larger lumps would not have their interior raised fully to the boiling temperature.[14] Roberts suggested that just prior to sealing the tubes, a reflux of outside air might bring in germs. More important, he thought splattering within the boiling mixture might cause some material dried near the top of the flask walls to avoid reaching the full boiling temperature.[15] On 16 April 1874 a long paper by Roberts was read at the Royal Society of London, offering many detailed experiments in support of the idea that only the most extreme measures could guarantee that infusions were not contaminated by air germs and guarantee that the infusions were not prevented from being exposed to the full boiling temperature.[16] Roberts found, for instance, that infusions containing fragments of green vegetables could be boiled twenty minutes or more without being sterile. However, if the vegetables were pulverized thoroughly in a mortar beforehand, sterilization became easier, suggesting to Roberts that the difficulty might be due to "some peculiarity of surface, perhaps their smooth glistening epidermis, which prevented complete wetting of their surfaces."[17] Boiled egg white showed similar difficulties, which he attributed to its alkaline pH, as Pasteur had found that milk and other alkaline substances were much more difficult to sterilize by boiling than acidic ones. Most difficult of all to sterilize, Roberts found, was hay-infusion neutralized or slightly alkalized by the addition of ammonia or liquor potassæ. This required at least one or two hours of boiling to be rendered sterile. Indeed, in some cases bulbs boiled for three hours "nevertheless became turbid and covered with a film in a few days."[18] This led Roberts to conclude that in some rare circumstances current-day abiogenesis might actually be occurring; though he still insisted, "If future investigations should establish the occurrence of abiogenesis, this would not overturn the panspermic [air-germ] theory, it would only limit its universality; and it may be predicted with some confidence that if abiogenesis exists the conditions of its occurrence can only be determined by an inquirer who is fully alive to the truth and penetrating consequences of the panspermic theory."[19]

This would clearly not include Dr. Bastian. He remained convinced that invoking air germs nobody had been able to see (including Roberts) was tantamount to begging the entire question. To *assume* the existence of air germs because one felt spontaneous generation to be impossible was indeed to debate preconceived assumptions rather than factual evidence.

The Germ Theory of Disease

Pasteur had suggested from the time of his initial work on spontaneous generation that growth of microbes causing the putrefaction of infusions was analogous to the process by which microbial infection also caused diseases in plants and animals. The progress of the germ theory of disease came to be bound up with spontaneous generation debates, since opponents of a germ theory could see spontaneously generated microbes as mere by-products of the disease process, not its cause. The German chemist Justus von Liebig opposed Pasteur on the germ theory. Liebig argued, like Pasteur, that the process of disease was analogous to putrefaction. But Liebig argued that fermentation and putrefaction were chemical processes, set in motion not by a living microbe, but by contact with a chemical catalyst, what in Britain was termed a "zyme." Thus, for Liebig and his followers communicable, or "zymotic," diseases were caused by contact with some tiny amount of a chemical poison, not a living germ.[20] As mentioned previously, Lister and Tyndall in Britain both began to follow up on Pasteur's idea. In addition, Burdon Sanderson had studied the British cattle plague of 1865–66 and concluded that the infectious particles, like those of vaccine lymph, were what he first called "microzymes" and later "bacteria."[21] In spite of this discovery, and in spite of the fact that Pasteur had already successfully challenged Liebig's notion that fermentation was a purely chemical process, Bastian was a follower of Liebig's theory, as were a number of other prominent British doctors and sanitarians, such as William Farr and Benjamin Ward Richardson. Bastian believed that the microbes found in the blood and tissues of diseased patients were by-products of the disease process, rather than its cause.

After his celebrated lecture, "Dust and Disease," John Tyndall wrote a letter to the London *Times* to emphasize the implications of his lecture and to convince the skeptical British doctors that the surgeon Lister was the only one who understood that the germ theory of disease was the key to conquering infectious disease.[22] It was not surprising that Bastian responded with a letter defending the popular theory of Liebig: "The question is . . . one which pertains most to the biologist and to the physician. It is, moreover, one of so complicated a nature that little save amazement will be excited in the minds of those conversant with all the difficulties of the problem, that Professor Tyndall should place so much reliance upon indirect evidence toward its solution, and should step forward on the strength of this, with the view of establishing a doubtful theory of disease, to which he, by his own confession, has so recently become a convert."[23]

Bastian urged that the settlement of the spontaneous generation debate would greatly help in determining whether the germ theory of disease had any value. He furthermore expressed irritation at Tyndall's blunt claim that only physicists and chemists approached things from a scientific point of view. Tyndall for his part was also taken aback, apparently not realizing that his advice to doctors would be taken to be so heavy-handed. His reply to Bastian escalated the personal tone of the dispute. He also expressed surprise that any modern man of science could remain unconvinced by Pasteur's experiments on spontaneous generation: "Dr. Bastian refers to these experiments as 'ambiguous.' As I perused his statement the words of an honest Catholic writer . . . came to my mind. 'There is,' said he, 'more sound theology in one page of Luther than in all the writings of Thomas Aquinas.' In like manner I would say that there is more solid science in one paper of Pasteur than in all the volumes and essays that have been written against him."[24]

Bastian replied that "it was precisely because I thought the theory of putrefaction which Professor Tyndall deems 'beyond dispute' was in reality utterly erroneous, that I ventured to caution him that . . . [he has] been standing on treacherous ground." In support of this opinion, he cited

> recent experiments made in concert with Professor Frankland. These results showed that living things were contained in the solutions of . . . hermetically sealed flasks, which . . . had been submitted for a period of four hours to a temperature varying from 148 deg. Centigrade to 152 deg. Centigrade. And seeing that there is not one tittle of evidence to show that living things can withstand . . . such a temperature for a single instant, . . . I have, I think, every right to conclude that the living things . . . had been produced *de novo,* and without the agency of "germs," by changes arising in the solutions after the flasks and their contents had been exposed to this very high temperature.[25]

Because of this connection, the progress of the germ theory of disease came to be bound up with spontaneous generation debates. The success of spontaneous generation experiments greatly increased skepticism toward the germ theory, especially among British doctors. This was highly frustrating to Tyndall and other supporters of the germ theory.

The Role of Heat-Resistant Spores

In addition to Lankester's and Roberts's criticisms of Bastian's experiments, John Tyndall tried to show that if the infusions were boiled in an atmosphere absolutely free of dust, they would never show growth of microbes. Pasteur and Lankester even suggested that some microbes must be able to survive temperatures of 100°C or higher. But as long as the existence of heat-resistant spores remained unknown, and such resistance thus unproven, there would always be cases in which believers in spontaneous generation could sincerely believe it was occurring, such as the case of Roberts's hay infusion. Even John Tyndall, one of Bastian's most ardent opponents, was baffled from October 1876 until early January 1877 when he could no longer produce infusions barren of microbial life, after having done so for many months in a challenge to Bastian.

The discovery of heat-resistant bacterial spores was first made by the German botanist Ferdinand Cohn and published in July 1876.[26] Tyndall established the role of spores in experimental infusions shortly thereafter. Cohn had met Tyndall on a visit to London in late September 1876 and given Tyndall a copy of the latest issue of his journal, *Beiträge zur Biologie der Pflanzen,* in which Cohn described the discovery of heat-resistant spores, especially in the hay bacillus. Thus, when Tyndall began to find that he could no longer successfully sterilize his infusions, he began to suspect a bale of hay recently brought into his laboratory at the Royal Institution as the possible cause. He checked all of his dust-free cabinets for leaks and repeatedly tried to obtain boiled infusions that would remain sterile in them. All his precautions to keep the air dust-free were to no avail, showing that dust was not the only important factor in spontaneous generation experiments (and that perhaps Bastian's work need not have been careless in that regard). Only when he finally removed his experiments to Kew Gardens and was able to obtain sterile boiled infusions again did Tyndall consider it proven that spores were involved.[27] He then tried to find a way to kill off the spores. By February of 1877 he could report that a new technique, fractional sterilization, was successful, even with spores that had been able to survive *hours* of boiling.[28] Tyndall found that boiling the solution for only a minute, then successively repeating that brief boiling after waiting for the solution to cool, provided sterility even in the most obdurate hay infusions. At most four or five repetitions were required. This seemed to show once and for all that heat-resistant spores had been the culprits producing growth in many of the otherwise carefully prepared infusions that had been considered instances of spontaneous generation.

Thus, in the end, the only truly decisive factor was the discovery of those spores. Once that was accomplished, Bastian was immediately sunk. His (or even Burdon Sanderson's) skill as an experimenter became an issue beside the point, and there was no longer any need to doubt what Huxley had recognized from the beginning: that current-day spontaneous generation was an illusion.

This, at least, is the standard story of the spontaneous generation debate in Britain. There is, however, another version of this story to be told. That more complicated and more interesting story will be the subject of the remainder of this book.

Major Victorian Scientists through the 1860s

Before going further, it will be useful to pull back and describe the major London scientific institutions and the careers of some of the key figures involved. In the 1830s and 1840s the Royal Society, the Royal Institution, and the Hunterian Museum of the Royal College of Surgeons were among the most important scientific institutions.[29] These were accessible almost exclusively to the well to do. Indeed the Royal Society underwent a series of reforms after receiving criticism from the new BAAS (founded in 1831) for electing members with strong aristocratic ties, but little scientific talent. The great Whig Reform Act of 1832 signaled the beginning of a gradual thawing of social institutions as a whole, for example by extending the franchise to non-Anglicans (Dissenters) for the first time. Still, accessibility of the most prestigious institutions of science to all was only beginning to be a reality by the late 1850s.

The Royal Institution had gained fame through the chemical researches of Humphry Davy and the work on electricity and magnetism of Davy's student Michael Faraday.[30] Faraday was a dominant authority in British science by the 1840s and 1850s. He was greatly concerned with increasing the public's understanding of science and had expanded the Royal Institution's public lectures on science, including instigating an immensely popular Christmas lecture series for children. The physicist John Tyndall was a newcomer to the Royal Institution in the 1850s, but quickly became a well-known popularizer of and public spokesman for science on the London scene. He shared Faraday's deep concern for the public's susceptibility to pseudoscience and quack medicine. Tyndall was known for confrontational tactics, such as challenging Anglican divines in 1866 and after to engage in a scientific test of the efficacy of prayer.

Some of the scientific societies, especially the Geological and the Linnean Societies, were also well established by this period.[31] By the 1850s numerous new societies had sprung up, though with a degree of influence and authority commensurate with their youth. These included the British Medical Association (founded in 1832 under a different name),[32] the Statistical Society, and the Ethnological Society. The new universities, the Whig University College and the Tory Kings College, had become solidly established as well, though they had decidedly different reputations (liberal, even radical vs. conservative, respectively) regarding the kind of scientific ideas researched and taught within their walls. University College London (UCL) had been perceived, especially in the late 1820s and early 1830s, as a place of radical ideas, as exemplified by its professor of zoology and comparative anatomy Robert Grant, an outspoken Lamarckian evolutionist and proponent of French philosophical anatomy. Grant was a favorite of the medical reform agitator Thomas Wakley and his journal the *Lancet*.[33] Grant had been Darwin's first mentor in science at Edinburgh in 1827. From him Darwin had imbibed the evolutionary ideas of Lamarck and the philosophical anatomy of Geoffroy Saint-Hilaire. The latter doctrine held that animal structure was determined by homologies with fundamental underlying unities of form, rather than primarily dictated by the needs of function as the dominant British tradition of natural theology held. The origin of these homologies, of the underlying unities of form, was not known. All through the 1830s and early 1840s Grant espoused these ideas in his teaching at University College and from his respected position at the Zoological Society of London.

Despite the initially radical image that figures like Grant brought, many of the professors attracted to "the Godless college on Gower Street" in the 1830s were equally talented scientists, and they eschewed extreme versions of democratic politics or medical reform. William Sharpey in physiology and Thomas Graham in chemistry, in addition to Grant, soon gave the college a very strong reputation for science. Grant had no significant British rival in comparative anatomy until the late 1830s, when he came into conflict and competition with a younger but rapidly rising talent at the Hunterian Museum, Richard Owen. University College continued to be Grant's main base of operations until his death in 1874. Adrian Desmond and James Moore have convincingly shown that by the 1850s such ideas as evolution, like University College itself, had ceased to have the extremely radical appearance they had twenty years earlier.[34] The politically explosive 1830s and "the hungry forties" were over, calmed by more than two decades of Whig reforms and the ascendancy of a new industrial middle class. Darwin's Malthusian version of species change, published in his *On the Origin of Species* in 1859, had helped to create a new image of evolution as a respectable competitive enterprise, mirroring Whig liberals' and successful industrial entrepreneurs' view of society. This was far more palatable in Britain than Lamarckian transmutation, associated as that earlier doctrine was with the time and politics of Revolutionary France. We shall look back in on Grant and the place of his ideas in chapter 3.

Before this change in public attitudes, however, Grant's chief rival Richard Owen became much more widely respected and more securely funded than Grant ever had; by the 1840s being called "the British Cuvier." Owen built his career by carefully tailoring his science in opposition to Grant's in order to cultivate and then to secure patronage from the Tory Anglican establishment. The established authorities saw Grant's evolutionary science as a threat to Christianity and to the aristocratic privileges of the established social order. Thus, up until the mid-1840s Owen was explicitly and vehemently opposed to all doctrines of species transmutation and of spontaneous generation. By 1845, however, Owen became more and more interested in the tradition of philosophical anatomy. Owen had studied this approach as far back as the late 1820s, but had kept silent about his interest because the emphasis of philosophical anatomy on form was at odds with the dominant natural theology. In natural theology, a Creator designed all aspects of a living thing in deliberate subservience to the function it must perform.[35] Owen had closely followed the work of both Étienne Geoffroy Saint-Hilaire on homologies of form and of Georges Cuvier on the dominance of function; he studied under both men in Paris at the time of famous public debates between the two in the French Academy of Sciences.[36] Owen had emphasized work on function in opposition to Grant from 1830 to 1845 because of the appeal of that approach to his Tory Anglican patrons.[37] Although Grant was still publishing Lamarckian ideas of evolution and spontaneous generation as late as 1861,[38] he had been marginalized among "respectable" London scientists by 1850 or so.

Now more solidly established, with a scientific reputation that would pass conservative and religious litmus tests, Owen could afford to work out some mixture of the role of homologies of form with functional anatomy. He became interested in the work of the German biologist and *Naturphilosoph* Lorenz Oken, even more speculative in its exploration of homology than that of Geoffroy. Indeed, in 1847 Owen orchestrated the publication by the Ray Society of the first English translation of Oken's *Lehrbuch der Naturphilosophie*. This work was highly speculative, even going so far as to support spontaneous generation of all life from "primordial slime (Urschleim) vesicles" in the sea. When the stodgy council of the Ray Society realized, after the book was already in print, that it contained such heresies against natural theology, an extremely vituperative debate broke out in the Society, assessing blame and enforcing new review measures to prevent any such embarrassing oversight in the future.[39]

Owen's interest in the 1850s moved toward some kind of natural laws that would drive a process of inherently progressive species transmutation, though he did not dare bring such ideas out in print. Then in 1858 and 1859 he was abruptly forced to show his hand by Darwin's publication of the theory of natural selection. The Darwinians (and many historians since) mistakenly interpreted Owen's response as a jealous attempt to claim credit for an idea he could never have believed prior to 1859. Rupke has suggested that the rift between Owen and the Darwinians preceded the *Origin* and was actually rooted in Darwin and Huxley's attempts to block Owen's most ambitious project, creation of the British Museum of Natural History. They did this, says Rupke, because of their desire to curtail Owen's power monopoly over life sciences in London.[40] At least some of the attempts on both sides to exaggerate their theoretical differences may have been motivated by personal hostility and professional competition as much as by any gulf between their ideas. At any rate, it does seem clear that Owen, by the mid-1850s at the latest, was a believer in evolution, though by a process quite different from natural selection.

Thomas Henry Huxley was a generation younger than Owen, but by the 1850s he was a fast-rising star on the scene of metropolitan science. More significant than a difference in age, Huxley, like John Tyndall, Herbert Spencer, and a number of other aspiring scientific talents, came from a modest background with no connections. He was fighting for jobs in science for young professionals based on talent alone. During the late 1840s and most of the 1850s this meant a difficult, overworked, hand-to-mouth existence producing barely enough income to support a wife and family. Thus, this new "Young Guard" of scientists (as Roy MacLeod has dubbed them) worked hard to make science a *profession* and to create jobs teaching science and doing research. They very jealously eyed Owen's success at concentrating in his own hands a huge share of the limited resources available for scientific research. In particular, Owen had monopolized the fields of biology and paleontology—exactly those fields in which Huxley specialized. So, just as Owen had worked furiously to edge out the older Grant in the 1830s and 1840s, Huxley labored incessantly to compete with Owen. Huxley seized on Darwin's theory and publicly defended evolution, greatly increasing his visibility, especially with the educated public. As a result, in a series of acrimonious debates between Huxley and Owen during the 1860s, many came to perceive Owen as opposing evolution and supporting special creation, when in reality he opposed only natural selection as the means of evolution.

It was in 1864 that a handful of the "Young Guard" scientists and their sympathizers first began to formally organize, calling themselves the X Club. The X Club began as an informal group of friends who met monthly to discuss professionalizing science and advancing Darwin's cause, among other things. The nine members met regularly for over twenty-five years and exercised enormous influence on the development of British science and science education. In general they supported each other's individual initiatives, and they held many important scientific offices, notably in the Royal Society of London.[41] The members included Huxley, Tyndall, Spencer, botanist Joseph Hooker, chemist Edward Frankland, UCL math professor Thomas Hirst, publisher William Spottiswoode, banker John Lubbock (whose interests included entomology and anthropology), and zoologist George Busk.

Among the X Club, second only to Huxley in public visibility as a spokesman and popularizer of science, was John Tyndall, professor of physics (natural philosophy) since 1853 at the Royal Institution.[42] Tyndall had become well-known for public lectures popularizing the latest scientific discoveries on everything from glaciers, to the reality of atoms, to conservation of energy, to why the sky is blue. In addition to showmanship he had, like Huxley, come from a relatively modest social background and had advanced to a prestigious post by sheer determination and hard work, engaging in highly public and vitriolic controversies with scientific opponents along the way. Not surprisingly, the two men had become fast friends by the 1850s. Tyndall had the hotter temper; when the Anglican Church declared a national day of prayer to combat the cattle plague epidemic in 1866 and to save the Prince of Wales from a bout of typhoid fever in 1873, Tyndall angrily challenged the Church to a scientific experiment to test the efficacy of prayer. He also spoke out constantly against the large number of sectarian doctors he considered quacks. Because his field was physics rather than anatomy or paleontology, Tyndall did not immediately come into conflict with Richard Owen, as the other X Club Darwinians did. Only in the early 1870s did his personal and professional relations with Owen sour.

2

"Molecular" Theories and the Conversion of Owen and Bennett

During the 1850s and 1860s two figures of very great stature and influence, both established in centers of scientific respectability, did change their natural theology views and argue for the existence of spontaneous generation in their writings. In this chapter I will explain how the mainstay of conservative anatomy, Richard Owen, and a respectable Edinburgh professor of medicine and physiology, John Hughes Bennett, could come to take a stance so novel in the British context. I argue further that the specific circumstances of Owen's and Bennett's conversion were crucial in setting the stage for a decade (1868–1878) in which spontaneous generation even became, for once, quite popular in Britain. The conversion of these two powerful scientists to advocacy of spontaneous generation influenced British thought on the issue in a measure equal to the influence of Pasteur's experiments until at least the early 1870s.

Brownian Movement and Histological Molecules

In order to understand the debates about spontaneous generation after the publication of *On the Origin of Species,* it is first necessary to trace the path of British science back at least to the year 1827. In that year the botanist Robert Brown first discovered "active molecules."[1] Here and there historians have noted in passing that many observers at first believed Brown's discovery supported spontaneous generation; however, it is commonly stated that Brown corrected this misunderstanding almost immediately in 1829 and that any linkage between the two phenomena came to an end shortly thereafter.[2] Here I will show that this is incorrect. Brownian movement and the "molecules" continued to be bound to ideas (both supporting and opposing) spontaneous generation up through the debates of the 1870s. Indeed, the final arrival at consensus that Brownian movement is a purely physical, nonvital phenomenon was *part* of those debates.

Much of the reason why this connection has been overlooked has to do with the use of the term "molecules." It has occasionally been noticed by historians that this term was used in a rather vague, loose sense by life scientists up through the late 1860s or early 1870s.[3] Some actors also noted this at the time.[4] It has been very little noticed, however, that amid this confusion there was at least one more or less coherent usage of "molecules" that had a clear meaning, albeit one quite different from the "molecules" of the chemists. John Hughes Bennett most clearly defined this usage and explicitly differentiated it from that of the physical sciences: "molecules" for the student of cells and tissues were microscopically visible particles with a diameter anywhere from about 1–2 μm or smaller, down to the limit of microscopic visibility and beyond.[5] They were found in the blood and in tissues, even within cells.

These "histological molecules," as I call them, were also noted by Theodor Schwann in his description of "free-cell formation" from cytoblastema, and by William Addison in the interior of white blood cells.[6] Others observing these particles under the microscope called them "granules," including Richard Owen, Charles Darwin, Joseph Jackson Lister, Lionel Beale, Albert Kölliker, and Joseph Lister.[7] In most cases these granules and molecules were understood to be capable of acting as reproductive propagules or at least of clustering together, in Schwann's sense, to form new cells.[8] Furthermore, at least some microscopists argued that the "free-cell formation" they, like Schwann, observed was in fact the process of heterogenesis, much as Buffon had interpreted his "organic molecules." These included Karl von Nägeli, Hugo von Mohl, Endlicher and Unger, Ludwig Büchner, Pouchet, Owen, and finally in 1868, Bennett. This notion was hardly novel: it had been noted before, both that Schleiden and Schwann's "free-cell formation" is in some ways the intellectual descendant of Oken's "Urzeugung" via primordial slime ("Urschleim") vesicles,[9] and Brown's active molecules were perceived by many of his readers to be a revival of a Buffon-like claim of spontaneous generation.

Notwithstanding historians' notice of these issues, the big picture of the "histological molecules" doctrine has not been recognized. Nor has anybody noticed the doctrine's implied support for heterogenesis.[10] By the 1860s a great many researchers in both medicine and laboratory biology saw these tiny objects as related to the vast problem-complex of generation and heredity, as well as to explaining disease, both contagious and degenerative. Because John Hughes Bennett was one of the most widely known teachers of physiology in Britain, his "molecular theory" of disease and of cell structure had perhaps the greatest impact. In January 1868 this impact was extended when Bennett stated explicitly that his theory also proved the reality of heterogenesis. Richard Owen, Britain's premier comparative anatomist, enthusiastically agreed with Bennett a few months later, in November 1868. Thus, the story of how "molecules" developed into a major research agenda is a crucial part of understanding how the doctrine of spontaneous generation came to have a heyday in Britain at this time. When "molecular theories" fell rapidly out of favor between 1869 and 1875, their demise removed a considerable pillar of support from spontaneous generation, above and beyond the frequently cited infusion experiments of the 1870s debate.

Owen's Role in Developing Ideas on Spontaneous Generation

The developing ideas of Richard Owen from the 1830s through 1880 on "molecules" and spontaneous generation are particularly revealing. This is especially true because of Owen's self-conscious comparison of his ideas in 1868 with those of Darwin and with Huxley's protoplasmic theory. Both Darwin and Owen had held an active interest in the role of "granules" or "molecules" in generation since the 1830s,[11] and continued to develop their ideas on this subject actively, culminating in Darwin's 1868 theory of pangenesis and an extensive critique of it by Owen. Darwin's pangenesis hypothesis was published in *Variation of Animals and Plants Under Domestication,* and within a few months, Owen published a critique and a rival theory, which had been printed only anonymously before then, in volume 3 of his own *Anatomy of the Vertebrates.*[12]

Darwin avoided the spontaneous generation question and attempted to be even-handed in discussing his theories about whether his "gemmules" also took part in a process of cell-formation from cytoblastema, describing the "omnis cellula" viewpoint as well as the cytoblastema view of the minority opponents.[13] He certainly attempted to place his own gemmules within mainstream opinion, however. Owen, by contrast, attempted to connect the mainstream position with the older, outdated and ridiculed preformation theory, and thus to discredit "those who still hold by this rag of 'pre-existence of germs,'" among whom he explicitly singled out Cuvier, Darwin, and Pasteur for criticism. Owen said that Pasteur's ubiquitous "germs" of bacteria were nothing but the old preformed "germs" in the egg of the preformationists. Thus, he argued, accepting the demise of the old pre-existence of germs theory implied that one *must* admit spontaneous generation as "the origin of single-celled organisms."[14]

But how could the former champion of conservative science, in the service of the Tory Anglican establishment, have come to argue for the inevitability of the radical doctrine of spontaneous generation? Owen's turn toward the work of Oken in the 1840s made spontaneous generation fit within a larger evolutionary framework that Owen was privately developing, as Evelleen Richards has shown.[15] Nicolaas Rupke has argued, further, that spontaneous generation was used by Owen after 1859 as an important way to distinguish his development theory from that of Darwin.[16] A much more detailed picture follows of the specifics of Owen's theorizing on spontaneous generation. It and other generation theories at the time were not merely trying to fit spontaneous generation into a purely theoretical context of transmutationism. These theories were also attempts to grapple with actual microscopic observations of "molecules" and the dialogue about their significance begun by Brown in 1828. Because Owen's conversion to spontaneous generation by 1858 also brought him into agreement with his old arch-rival Robert Grant,[17] it must represent much more than just a strategic marker for distancing himself from the Darwinians. Spontaneous generation played an important role in Owen's overall evolutionary theorizing and in his views on generation theory, which had been percolating since the late 1830s.

Owen's interest in "granules" can be documented as early as 8 March 1836,[18] while Darwin's began much earlier, as a result of his studies in Edinburgh with Robert Grant.[19] Owen focused his attention more on the infusorial monad (protozoan) as the most likely analog of the "germ" from which an entire organism could develop, relying on Johannes Müller's (1837) theory about a fixed amount of vital force immanent within the germ.[20] Darwin's attention, by contrast, had been directed from his Edinburgh days to the much tinier granular matter *within* such larger units (cells), as that which might be the "germs" from which reproduction of simple organisms, up through zoophytes, proceeded. Darwin was stimulated and confirmed in this view by Robert Brown's 1828 discovery of "active molecules," and sought Brown's advice for continuing his investigations. He did not at first believe what was taken to be Brown's claim that the molecules were self-active, preferring Henslow and Brongniart's notion that their vital force was superadded from without. This view ruled out the ability of such particles to lead to spontaneous generation in the way Buffon and Needham had originally suggested.[21]

This distinction is crucial: those who saw Brownian molecules as self-active believed the particles capable of spontaneous generation. The radical medical reformers that Adrian Desmond has studied also saw them as the metaphor in nature for "bottom-up," democratic social and political reforms.[22] Those who opposed spontaneous generation were adamant that the motive force had some external, purely physical source such as evaporation, uneven heat distribution in the fluid, electricity, and so on. It is worth emphasizing that the majority of scientists had come to accept this view by the time of Brown's death in 1858, even though nobody up to that point had yet produced a satisfactory explanation for an external physical cause of what was by then being called "Brownian movement."[23]

By early 1837, however, in his transmutation notebooks, Darwin began to speak of the granules as "living atoms," reflecting a theoretical shift on his part.[24] Furthermore, in July of that year an article by the Berlin microscopist Christian Ehrenberg first appeared in English translation. In this influential article, Ehrenberg claimed that painstaking microscopic study had shown that the infusorial monads did not arise by spontaneous generation, but rather from those much smaller granules that he assumed to be "eggs" of the infusorians.[25] This seemed to prove that it was not the protozoan "monads," but rather the tiny granules they contained that were the real ultimate living particles, capable of reproduction.[26]

Darwin's thinking on granular matter and its fundamental role in generation progressed. He thought this implied unity of all living matter (in a way that at first strongly suggested a possible manner for the transformism of species). In the end the granular matter faded into the background in Darwin's writings after his Malthusian insight of September 1838 and never appeared essential in any written version of his transmutation theory. These granules remained important in Darwin's thought throughout the period leading up to *On the Origin of Species*, however, and finally emerged into public view as the "gemmules" of his pangenesis theory of generation and inheritance in the 1860s.[27]

Historians have not followed the changing ideas of Owen on these granules through the decades after 1837. But Owen's writings went through a major shift in this domain, in a way related to his much broader theorizing about philosophical anatomy and spontaneous generation as promulgated by Lorenz Oken.[28] Yet, initially Owen's interest in granular matter was embedded (in public at least) in a context that opposed Lamarckism and those other doctrines,[29] up through at least the mid-1840s. Then, by the late 1840s, Owen developed a private interest in the ideas of the German *Naturphilosph* Oken. Owen had first heard such ideas from his teacher Joseph Henry Green and then again in 1830 in Paris during the Cuvier-Geoffroy debates. Stimulated by his reading of Oken, by the early 1850s Owen finally began to publicly express his earlier musings over philosophical anatomy. Like much of the biological community by this time, Owen was reevaluating the importance of Geoffroy's ideas, and his comparative anatomy had taken on a strongly Oken-like transcendental flavor.[30] Because of the dominance of the Darwinians' accounts in the history of evolutionary ideas, almost all of Owen's ideas from this point forward, including those on spontaneous generation, have until recently been regarded as religiously biased and of little scientific influence after Huxley's victories over Owen in the early 1860s.

By early 1837, during the period of close contact between Darwin and Owen, the threat of spontaneous generation ideas was much on the minds of British naturalists and medical men. Lyell had savagely attacked Lamarck on this point, so fundamentally incompatible with his uniformitarianism as well as with British natural theology. Robert Grant was teaching Lamarckian spontaneous generation in his zoology course at University College London, and radical medical reformers were actively promoting it as an underpinning for bottom-up democratic political agitation, though being loudly denounced by the Oxbridge dons of science.[31]

In the first two months of 1837, newspaper headlines were splashed with debate and recriminations about "the most famous experiment of the first half of the nineteenth century," the claim by Andrew Crosse to have proven spontaneous generation of insects by an electrical experiment.[32] Few could be oblivious to the sensation created all over Britain by the debate over Crosse's experiments and their anti-theological implications, but Owen had additional reason to take detailed notice of Crosse's scientific claims. On 10 February 1837, Crosse wrote to Owen sending him specimens of the allegedly spontaneously generated insects, presumably trying to enlist Owen's zoological authority. Owen must have shown at least some initial courtesy, for, by April of that year, Crosse had written again, this time with an extensive account of his experiments and the growth of one of the insects he had dubbed *Acarus crosii*.[33] The vast majority of the natural theology–oriented British scientific community ridiculed Crosse. Michael Faraday, who had argued in 1829 that respectable Robert Brown had only been misunderstood by those who thought he was advocating self-active matter in his announcement of "active molecules," had been cited in some press accounts as supporting Crosse. Faraday went out of his way to publicly insist that he held no such opinion.[34] Faraday was by no means alone in this stance. Thus, fully aware of the current claims for spontaneous generation in Britain from Grant and Crosse, Owen was also aware that such claims were totally unacceptable to his patrons, including Adam Sedgwick and William Whewell, during this vulnerable early phase in his career. By the time of the second in his series of Hunterian Lectures, on 4 May 1837, Owen took a public stand against spontaneous generation, citing Ehrenberg's work as proof.[35] Owen was building his patronage network at this time among established conservative men of science by providing a version of anatomy and physiology that specifically opposed the transcendentalist "French" anatomy being taught by Robert Grant.

It is clear from his 1840 Hunterian Lectures, however, that Owen remained keenly interested in the question of spontaneous generation and kept current with all new opinion and experimental work on the subject. In that later series of lectures devoted entirely to the subject of "Generation," Owen's first two lectures were mostly on the history of spontaneous generation theories and evidence for and against them.[36] And while his publicly stated conclusion was still against the doctrine, he dwelt at considerable length on evidence in its favor from Oken and Karl Friedrich Burdach. Furthermore, in the final draft, he elaborated upon the opinion of Dr. Allen Thomson of Edinburgh "in his recent . . . article 'Generation' in the Cyclopedia of Anatomy and Physiology—among the latest and best summaries . . . of that function which has been published in this country." Thomson, stated Owen, "considers that the balance of evidence is in favor of the Spontaneous production of Infusorial animals, vegetable mould, and the like."[37]

Through the 1840s, as Owen's views on the subject developed, the politically contentious climate in Britain only worsened. Therefore, even if he found Thomson's or Oken's ideas appealing and thought-provoking (Oken's "primordial slime vesicles"[38] were after all evocative of the granules that Owen had been interested in since the 1830s), and ventured some tentative opinions on transcendental anatomy, the debacle of the Crosse incident, followed in 1844 by a similar uproar over *Vestiges of the Natural History of Creation* made it clear to Owen that spontaneous generation and transmutation of species were still taboo in official science circles. Though *Vestiges* was widely read and discussed by middle-class audiences, the authorities in the sciences harshly criticized the book from top to bottom. Such tentative balloons as Owen *was* willing to float during the 1840s received clear negative warnings in response from Sedgwick and the *Manchester Spectator.*[39] And Sedgwick's attack on the spontaneous generation ideas found in *Vestiges* was merely reflective of the entire elite scientific community.

John Hughes Bennett and "Histological Molecules"

During this time the importance of "granules" and "molecules" in generation, disease, and other physiological processes continued to be brought to Owen's attention; for example, through the works of Schwann on cell theory and of Addison on blood corpuscles. Schwann claimed that cells were the essential structural and functional elements of both plants and animals, but that new cells formed by the accumulation of "molecules" from a fluid blastema in a crystallization process.[40] This corresponded to popular theories in pathology, such as those of Karl Rokitanski and John Hughes Bennett.[41] Bennett's early cytoblastema ideas continued to develop and spread widely because of his reputation as professor of the Institutes of Medicine (later equivalent to professor of physiology) at Edinburgh. In 1842 Addison reported the existence of "molecules," or microscopically visible granules, in the colorless blood corpuscles and free in the blood fluid from patients with various diseases. Seeing the same granules in epithelial cells throughout the body, Addison concluded that "all varieties of epithelial cells are formed by colorless blood corpuscles," and that "the granules in the colorless cells circulating in the blood, were reproductive units capable of becoming young cells after their discharge."[42]

Rudolph Virchow began to oppose the cytoblastema theory of cell origins in his 1855 essay, "Cellular Pathology," and in his 1858 book of the same name. Having at first supported cytoblastema theory through the 1840s like Bennett, Virchow swung to the opposite extreme, claiming rather dogmatically that cells never come from any source except the division of pre-existing cells. Bennett then became more strident in spreading his "molecular theory," in response to the spread of Virchow's *omnis cellula* doctrine. Bennett considered this an excessively broad claim for the cell theory and thought Virchow arrogant in making such sweeping pronouncements.

Historian L. J. Rather observed that Schleiden and Schwann's blastema theory of cell formation "did not disappear at once from the scene with the advent of Virchow's *omnis cellula a cellula,* continuing instead to flourish somewhat longer than is usually recognized by historians of science."[43] John Hughes Bennett's molecular theory is one of the descendants of the blastema theory cited by Rather, though he adds, "it should not be supposed that Bennett's views were simply the result of his longstanding antagonism to Virchow. They were shared by others" well into the 1870s, including University of Edinburgh anatomy professor William Turner and the noted French microscopist and histologist Charles Robin.[44] Bennett's antagonism to Virchow had originated in the late 1840s and early 1850s in a priority dispute over the discovery of leukemia. At the same time, however, Bennett had begun a critique of the cell theories of Schleiden and Schwann and of his Edinburgh colleague John Goodsir.[45] By 1852 he was proposing his own theory. Schleiden and Schwann's cytoblastema, or structureless formative fluid, was not observed to be the actual source of cells, he claimed. Yet in most cases, neither were other cells directly, as Goodsir had claimed in 1845 and Virchow in 1855. When Huxley wrote a review on cell theory in 1853, Bennett also found shortcomings with his version. The solution, he argued, was to focus on the primacy of units more basic than the cells: the "molecules" that Schwann had mentioned in his treatise of 1839.[46] By

1863 Bennett had developed his theory into the basis of an entire lecture course, in which he compared his "molecular physiology, pathology, and therapeutics" to the atomic theory of Dalton. The atomic theory, Bennett said, "has unquestionably given a great impulse to chemical science, but has done little for the science of organization. It has . . . thrown light on the proportional combinations of the chemical elements, but has taught us nothing whatever as to the development and growth of plants and animals."[47] And while Schleiden and Schwann's cell doctrine had done much to advance histological knowledge, Bennett insisted that it

> has yet led many minds to the conviction that it cannot embrace all the facts of organization. Hence it appears to me evident, that with a view to making further progress, . . . we must substitute for the hypothetical atoms of the chemist, the visible molecules of the histologist, and demonstrate how all research and discovery in recent times tend to support a molecular, rather than a cell-theory of organization. It will be my object . . . to blend the well-known doctrine of Schleiden and Schwann into a theory of wider application; to show how the facts of physiology give it the most unequivocal support; and lastly, indicate the manner in which it must constitute the basis of a sound therapeutics.[48]

He believed that normal cells, but also pathological formations such as tumors and tubercles, formed from these "histogenic" molecules and that those structures could all break down into "histolytic" molecules. A wide audience of researchers with biological and medical concerns immediately understood Bennett's reference to the "visible molecules of the histologist." These microscopic globules, much smaller than cells, had been seen by investigators dating back to Buffon and were known as "granules," "organic molecules," "living atoms," and numerous other names. They were widely believed in pathology to be the units out of which new pathological growths, or the cells in them, formed.[49] William Addison believed the "moving molecules in the interior of cells" in the blood were crucial to understanding the formation of those cells, as well as to processes of disease and inflammation.[50]

In the British context, it should be noted, these molecular theories were developed by men who almost all concluded that Brownian movement was caused by some nonvital external physical force, and that spontaneous generation was impossible. Bennett up through at least 1863 was clearly among these. He stated that "the molecular, therefore, is in no way opposed to a true cell theory of growth, but constitutes a wider generalization and a broader basis for its operations. Neither does it give any countenance to the doctrines of . . . spontaneous generation. It is not a fortuitous concourse of molecules that can give rise to a plant or animal, but only such a molecular mass as descends from parents."[51] After 1863, Bennett, like Owen before him, began to reconsider this denial of spontaneous generation. And by 1868 both men very publicly supported spontaneous generation by way of the clumping together of "molecules."

In 1843 Addison had distinguished the term "granule" for those objects visible as bright points surrounded by a dark circle and measuring up to as much as one-fourth or one-fifth the diameter of a red blood cell. He distinguished "molecules" as so small that no bright center could even be seen, downward to below the limit of visibility. Bennett essentially adopted this distinction in his writings on the subject, saying that "molecules" varied in size "from the four thousandth of an inch down to a scarcely visible point, which may be calculated at less than the twenty thousandth of an inch in diameter," though he noted that the distinction was purely arbitrary, relative to magnifications currently available: "In the same manner that the astronomer with his telescope resolves nebulae into clusters of stars. . ., so the histologist with his microscope magnifies molecules into granules, and sees further molecules come into view."[52] Bennett also argued that knowledge of the molecules could revolutionize therapeutics just as much as cell theory had. For instance, he pointed out that "the vitality of these minute structural elements being inherent in themselves, must convince medical men that the morbid changes which they originate are extra-vascular," and thus offer theoretical proof of the uselessness of bloodletting and other formerly popular remedies.[53]

Thus, Bennett's molecular theory was formulated as part of a long process of critical examination of cell theory and an even longer-standing discourse about ultimate subcellular particles of life, not as a specific objection to Virchow's *omnis cellula* doctrine. By the time Virchow's *Cellular Pathology* appeared, however, Bennett was used to scrapping,[54] and was convinced by the development of his own theoretical line that Virchow was not merely wrong, but arrogantly sweeping in his inaccuracy. He summed up his view of Virchow's campaign thus:

> Parodying the celebrated expression of Harvey, viz., *Omne animal ex ovo,* it has been attempted to formularize the law of development by

> the expression *omnis cellula e cellula,* and to maintain "that we must not transfer the seat of real action to any point beyond the cell." In the attempts which have been made to support this exclusive doctrine, and to give all the tissues and all vital properties a cell origin, the great importance of the molecular element . . . had been strangely overlooked. It becomes important, therefore, to show that real action, both physical and vital, may be seated in minute particles, or molecules much smaller than cells, and that we must obtain a knowledge of such action in these molecules if we desire to comprehend the laws of organization.[55]

Both Bennett and an anonymous reviewer of Virchow in the *British Medical Journal*[56] voiced skepticism about Virchow's claims in 1861, in addition to those critics cited previously.

After its initial presentation at the BAAS in 1855, Bennett's theory became widely known in Britain through the early 1860s. However, shortly after Bennett's lectures on the molecular theory began to appear serially in the *Lancet* in January 1863,[57] Lionel Beale, the renowned microscopist of Kings College, launched a frontal assault on Bennett's theory in a series of letters to the *British Medical Journal.*[58] Bennett replied to Beale in an exchange in the journal lasting until April, while his lectures continued appearing in the *Lancet* through the rest of 1863.[59] Beale's first point was to suggest that Bennett's denial of any countenance of spontaneous generation seemed to be contradicted by the apparent similarity between molecules of formation and molecules of disintegration, saying "surely he will not maintain that lifeless particles become aggregated together, and form a living mass."[60] Bennett replied that the disintegrated molecules from a decaying tubercle were coming from a structure already alive, and that he did not thus imply any transition from nonliving to living. Bennett had not changed his public stance on this as late as 1867,[61] and he only took on Beale's vitalism directly with a more physicalist account after Huxley and Owen had done so in 1868–69.[62] Nonetheless, Beale's critique foreshadows many of the themes that would be much more developed in his attacks on Huxley, Owen, and the protoplasmic theory.[63] At this time many in Britain took a lively interest in the debate in Paris between Pouchet, who claimed to have experimentally demonstrated heterogenesis, and Pasteur, who opposed Pouchet via experiments of his own. Beale's critique seems to have stimulated Bennett to realize that his theory might in fact imply something like

Pouchet's heterogenesis, as just at this time Bennett began to look into the Pasteur-Pouchet debate, and within a few months, to carry out experiments of his own to challenge the claims of Pasteur.[64]

Beale also charged Bennett with using only low powers of magnification, up to 250x. Bennett, however, had been an early pioneer in introducing the new achromatic microscopes into British histology in 1841, and had been upgrading his research instruments with the latest available improvements ever since. Thus he replied that although he had used some older illustrations in his paper in the *Lancet,* he was currently employing microscopes capable of 700–1,200x but found none of his basic conclusions altered by the new capabilities. Indeed, Bennett was among several that leveled the same charge against Pasteur's work over the next few years.[65] Beale also challenged Bennett's claim to have seen molecules organize into chains, then fuse to form independent, free-swimming vibrios. "Although I have often watched such minute organisms with powers magnifying from 700 to 3000 diameters, I have never been so fortunate as to see even two particles unite," he said, and he did not think anyone else ever had, either. Instead, Beale had observed that the molecules grew by absorption and then divided, so that their numbers multiplied.[66] Beale was greatly opposed to the idea that vibrios could form by such a heterogenetic process, and repeatedly suggested that Bennett could not have seen this.

Beale's most potent criticism, however, was that Bennett defined the term molecule to include so many different kinds of objects that he was confusing particles with totally different properties in size, shape, and composition. Beale found it difficult to believe that the molecules' similarity in size could be a more important feature than the differences between them. To this Bennett responded that "other elementary parts . . . possess equal differences. Cells also vary in size, origin, chemical composition, complexity and properties; yet they are still cells."[67] This was consistent with his overall critique of cell theory.

Owen's Change and the Darwinians

In addition to the histological "molecular" theories, chemistry was also producing theories of disease based upon the chemists' molecules. Liebig had drawn an analogy between his theory of fermentation and the causes of disease, saying that both were due to processes of "molecular excitation," catalytically propagating. Liebig had a great following in England, and much of English medical theory from the 1840s until the 1870s concentrated on such chemical "molecular theories" of disease,[68] with figures like Bastian explicitly arguing for the tight linkage of this and the spontaneous generation doctrine.

Like Bennett, Owen at first may have fit these molecular and "force" theories within his larger schema without necessarily accepting the possibility of spontaneous generation. He may have remained reserved or skeptical on this topic even after becoming a full-blown convert to Oken's transcendental anatomy, despite Oken's enthusiastic support of spontaneous generation. But in 1859, 1860, and again in January 1862, we see unequivocal evidence of Owen's conversion to the latter doctrine.[69] Owen there cites Pouchet's work in opposition to Pasteur, and one of the most important resources Owen drew upon in his conversion and in formulating his immanent-force "ultimate atoms" as agents of spontaneous generation was his friendly correspondence over many years with Pouchet, beginning in 1845.[70] Felix Archimede Pouchet of the Natural History Museum at Rouen initially contacted Owen, complimenting the Englishman on his work, and the two continued to exchange papers and to cite one another's work. On 1 Sept. 1860, for instance, Pouchet wrote to thank Owen for having favorably mentioned his work in Owen's book *Palaeontology*.[71] Owen continued to support Pouchet's position throughout the famous debates with Pasteur from 1860 until 1864. For example, in his paper "On the Aye Aye," Owen specifically praised Pouchet's work on "the . . . hypothesis . . . that (organisms) are coming into being, by aggregation of organic atoms, at all times and in all places," calling the Frenchman "able in refutation of objectors to Heterogeny, and full of resource in its support."[72] Jeffries Wyman, another long-time correspondent of Owen, seems to have responded to Owen's interest in Pouchet by conducting his first experiments on spontaneous generation at Harvard not long after reading Owen's Aye Aye paper.[73]

An anonymous reviewer of William Carpenter's text on foraminifera (certainly Owen), writing in the 28 March 1863 *Athenaeum*, took Carpenter to task for Darwinian views on the origin of these tiny marine microorganisms, and suggested that Carpenter demonstrated the "weakness" of the Darwinian hypothesis of natural selection and the superiority of a Lamarckian view, including

> heterogeneous production of the primitive types of organisms. Agreeably with this principle, we conceive that "particles of apparently homogeneous jelly" are now, as of old, being aggregated through the operation of existing interchangeable modes of force . . . The assumption . . . [seems preferable] that at the period when life became possible on the earth's surface, the conditions were sufficiently varied to permit the conversion of the general polaric into a specific organic mode of force to operate . . . resulting in a variety of the simplest forms of life, such as "monad," "mucor," "amoeba," "lichen-spore," &c. and that such conditions have continued to operate in the heterogeneous production of "organisms without organs" to the present day.[74]

Such a synthesis of Pouchet's heterogenesis and Oken's speculations about spontaneous generation resulting from varying polarity of forces was uniquely Owen's line of thought.[75] Owen's convoluted prose style was unmistakable to readers as well.

The review provoked a defense by Carpenter, who pointed out that such views "by most naturalists . . . will be regarded as a far more 'astounding hypothesis' than the one for which it is offered as a substitute."[76] Just a month later, in April 1863, a letter on Carpenter's behalf was also written by Darwin, who admitted that "your reviewer sneers with justice at my use [in *On the Origin of Species*] of the 'Pentateuchal terms,' 'of one primordial form into which life was first breathed': in a purely scientific work I ought perhaps not to have used such terms; but they well serve to confess that our ignorance is as profound on the origin of life as on the origin of force or matter."[77]

Almost simultaneously, Darwin wrote privately about this episode to Hooker: "Who would have ever thought of the old stupid *Athenaeum* taking to Oken-like transcendental philosophy written in Owenian style. It will be some time before we see 'slime, snot, or protoplasm' (what an elegant writer) generating a new animal. But I have long regretted that I truckled to public opinion, and used the Pentateuchal term of creation, by which I really meant 'appeared' by some wholly unknown process. It is mere rubbish thinking of the origin of life; one might as well think of the origin of matter."[78]

Darwin also encouraged George Bentham to challenge the reviewer in an upcoming lecture to the Linnean Society on 23 May 1863.[79] He reported a few days later to Hooker that "the mention of Pasteur by Mr. Bentham is in reference to the promulgation 'as it were *ex cathedra*,' of a theory of spontaneous generation by the reviewer of Dr. Carpenter . . . Mr. Bentham points out that in ignoring Pasteur's refutation of the supposed facts of spontaneous generation, the writer fails to act with 'that impartiality which every reviewer is supposed to possess.'"[80]

Hooker replied with disdain for the whole affair, warning Darwin that when supposed scientific experts aired their disagreements in public with such a personal tone, science overall was the loser in the eyes of the public. Hooker believed that answering Owen in public merely gave him the opportunity for "another skilful appeal to popular prejudice," which in this case would be considered "triumphant in the eyes of the public. Science will be much more respected if it keeps its discussions within its own circle."[81] Darwin belatedly agreed, saying "if I am ever such a fool again, have no mercy on me."[82] This issue was a major concern to Victorian scientists, especially in the aftermath of confusion over antagonistic testimony by "scientific experts" in some spectacular jury trials around this time.[83]

Huxley discussed the *Athenaeum* episode with his friend Rev. Charles Kingsley:

> I am glad you appreciate the rich absurdities of the new doctrine of spontogenesis. Against the doctrine of spontaneous generation in the abstract I have nothing to say. Indeed it is a necessary corollary from Darwin's views if legitimately carried out, and I think Owen smites him [Darwin] fairly for taking refuge in "Pentateuchal" phraseology when he ought to have done one of two things— a) give up the problem, b) admit the necessity of spontaneous generation. It is the very passage in Darwin's book to which, as he knows right well, I have always strongly objected. The x of science and the x of genesis are two different x's, and for any sake don't let us confuse them together.[84]

Meanwhile, Gilbert Child, in experiments at Oxford (reported to the Royal Society in early 1864), Jeffries Wyman's experiments in Boston, and those of Hermann Schaaffhausen at Bonn (reported in September 1864) continued to pursue the investigation of spontaneous generation explicitly because it seemed to them to be a very likely implication of Darwin's theory of evolution. Child argued for the linkage in an influential review article in the *British and Foreign Medico-Chirurgical Review* for July 1864 and again in his *Essays on Physiological Subjects*.[85] Soon after, the British medical press noted of Schaaffhausen:

> He is a staunch adherent of Darwin, and generalizes from this point of standing regardless of the consequences. To him, spontaneous generation is the only wanting link in the chain of facts which prove the unity and unchangeability of Nature. He has watched organic decomposing substances and found that they give rise to the lowest imaginable and the smallest visible germs . . . Here, in England, Schaaffhausen will find much sympathy in his endeavour to unravel the mystery of protogenesis. Have not Bristowe and Rainey shown, step by step, the spontaneous generation of the Trichina? And has not Professor Hughes Bennett of Edinburgh seen the fine particles fly together, "fall in" in generation order, and turn out a perfect living vibrio?[86]

Owen was also favorably impressed by Schaaffhausen's work. It is important to note that Darwin's attitude toward spontaneous generation softened considerably at various points over the next two decades, as Huxley's analysis in "The Physical Basis of Life" sank in. However, he only admitted this privately, and only in response to work done by the "right sort" of "careful" and "trustworthy" men. Darwin named among these men Wyman, Child, Huxley, and John Burdon Sanderson.[87]

No sooner did Owen publicly take a stand in favor of heterogenesis than the editor of the *Lancet* hailed the spontaneous generation theory as the new wave, soon to triumph, and cited Owen as the latest eminent authority converted to the coming "more scientific" trend.[88] In the eyes of many, Owen had remained the bastion of conservative respectability, still the antithesis of the radical medical tradition championed by the *Lancet*.[89] Indeed, the *Lancet* reviewer of 13 February 1869 likened Owen's position to Huxley's "Physical Basis," saying, "It is no new thing to find Mr. Huxley leaning to the heretical side, so this may not be thought very surprising. But the most striking symptom of the general tendency of opinion is the conversion of Professor Owen—that great champion of teleological orthodoxy in science—to views which to a good many worthy folk will seem perfectly awful."[90]

The *British Medical Journal* had a similar assessment: "The mass of evidence is so overwhelming, that even Professor Owen, whom none would accuse of previous leanings in this direction, has lately . . . boldly announced his assent to the doctrine of heterogeny."[91] Owen saw the experiments of Pouchet confirmed by those of Gilbert Child of Oxford, Jeffries Wyman of Harvard, and John Hughes Bennett.[92] Bennett's influential summary in 1868 of the case for heterogenesis was widely cited, especially when the article was reprinted in the January 1869 issue of *Popular Science Review.*[93] Wyman, though friendly with Owen, was also widely respected by the Darwinians and was considered a model of experimental caution among those who had come out in favor of evolution in America. In March of 1865 Wyman expressed considerable skepticism about Pasteur's claim to have settled the spontaneous generation question experimentally: "As to Pasteur, the results obtained by recent experiments both in France and Germany show that he has taken a ground which he cannot defend, when he says that all life ceases in organic fluids boiled for a few minutes and exposed to air purified by heat. The opposite results of others are too many to allow him to stand unimpeached on this score of refutation."[94]

In addition to Oken, Pouchet, and Wyman, another likely source for Owen's changed ideas was his avowed debt to Michael Faraday's 1852 paper, "On Lines of Magnetic Force, Their Definite Character and Their Distribution within a Magnet and Through Space."[95] In his musings about Oken's notion of polarity and its application to generation of organisms, Owen seems to have freely applied Faraday's discoveries about polarity in magnets and the relation of magnetism to electricity. We have already seen some of his 1863 speculations on polarity of forces and their sometime conversion into "organic" or "vital" force. In his 1868 treatment, he speculates on the origin of "the simplest living jelly (*Protogenes* of Haeckel) or sarcode (*Amoeba*)" by spontaneous generation and further develops the analogy between the vital force of this protoplasm and the magnetic force:

> Magnetic phenomena are sufficiently wonderful, exemplifying as they do, one of those subtle, interchangeable, may we not say "immaterial," modes of force which endows the metal with the power of attracting, selecting, and making to move a substance extraneous to itself. It is analogically conceivable that the same *Cause* which has endowed His world with power convertible into magnetic, electric, thermotic and other forms or modes of force, has also added the conditions of conversion into the vital mode. Nerve-force we know to be convertible into electric energy, . . . and from the electric force so induced, magnetic and other modes have been derived . . . The direction, then, in which may be anticipated the replies to the ultimate question, will be toward an admission of the originating and vitalizing of the primary jelly-speck or sarcode-granule, by the operation of a change of force forming part of the constitution of Kosmos.[96]

Anticipating charges that his view was fundamentally materialistic and therefore anti-religious, he defended himself by claiming that a Biblical miracle could be explained via "lines of force, as 'luminous undulations,'" emanating from "centres," some of which are tangible and some of which are not: "If a blade of metal could move itself to and fro in striving to cleave the space between excited electro-magnetic poles, and could tell us its sensations, they would be those of sawing its way through a substance like cheese; but there is no visible impediment: nor, were luminous undulations to vibrate from the hindrance as from the plane of force resisting the pressing finger, would the hindrance be less 'immaterial.' Similarly, if lines of thought-force were visible, the 'ghost' would not on that account be more 'material.'"[97] He continued, "I cannot feel that I know more about the matter by calling the 'centres of force,' 'material atoms' or 'immaterial points,' and am resigned to rest at a point beyond which Faraday did not see his way."[98]

It must not be forgotten that Owen's attempt to forge links between magnetism, electricity, and vital force occurred in the context of thirty years of conflict. Mesmerists, "electricians," and other sects of popularized science had been battling since the 1830s with the newly emerging professional scientists, who claimed the sole right to establish standards of proper scientific work. The Mesmerists and electricians had tried to argue for just such easy transitions among magnetism, "animal magnetism," electricity, and vital energy, the highest form of which was "nerve force." Michael Faraday and William Benjamin Carpenter were early champions of the cause of a professional science, untainted by the popularized, amateur aura that surrounded Mesmerists.[99] Darwin and his supporters who founded the X Club followed this more distinguished, respectable model of scientific behavior. They saw the Crosse spontaneous generation episode, atheists and medical radicals, and *Vestiges of the Natural History of Creation* as examples of all that was nonprofessional and "bad science" in the opposing approach. Thus, the kind of linkages Owen pursued practically guaranteed that his support for spontaneous generation would be seen by many of the professionalizers as a throwback to all that Victorian science ought to be escaping. Owen resolidified in the minds of Carpenter, Huxley, and Tyndall the old negative associations that spontaneous generation ideas had twenty and thirty years previously. The logical argument that spontaneous generation was necessary as the underpinning for an evolutionary world was a compelling one for many Darwinians, however, so they found themselves caught between two conflicting agendas. Gilbert Child, for instance, argued forcefully for the linkage between evolution and spontaneous generation, but he noted: "The maggots in Redi's days, and the Acarus Crosii in our own, have been . . . fatal to a fair consideration of its claims . . . If people will look for it [spontaneous generation] in the 'more highly developed forms,' if they expect to see an elephant or an acarus spontaneously produced, it is no wonder that the whole theory is laughed out of court."[100]

How Owen saw himself in the context of this larger conflict we do not know, but after 1868 his view of the relationships among magnetism, polarity, and life seems to have changed little. Owen wrote an 1880 letter to spontaneous generation supporter Henry Charlton Bastian, for instance, asking for any references Bastian might know to publications "prior to 1868, containing a comparison of the selective act of the magnet with that of the amoeban," so that Owen could set straight any omission of credit to others in the next edition of *Anatomy of the Vertebrates*. And in that context his description of "centres of force" seems little changed. He admitted here for the first time that it was consideration of the "molecules" of Brownian movement that *led* him to this conception, demonstrating unequivocally the connection in his mind over many years between Brown's discovery and the spontaneous generation question:

> Among the trains of thought thereto leading [i.e. to his magnet-amoeba analogy] was this: centres of motion = ultimate molecules, atoms, etc. of the (to our touch) densest of substances, diamond, e.g.; are ever changing their relative position, are in "perpetual motion." When groups of such "centres" are so small as to need high magnifying power to be recognizable by our visual sense, they exhibit such motility, or its effects, as in the case of the "Brownian molecules": they seem so full of so-called "life," dancing about like midges in a sunbeam.[101]

Thus, the connection between spontaneous generation and Brownian movement was evident to many who took an interest in either issue. Bastian, for example, considered this relationship; he had initiated the correspondence with Owen in 1876.[102] Similarly, Charles Letourneau in France, in his 1878 textbook, *Biology*, defended Pouchet's spontaneous generation experiments and argued that Brownian movement actually included a *range* of movements, "a kind of gradual transition from inorganic movement to organic motility."[103] He was suggesting that, within the types of motion labeled as "Brownian," there was a spectrum, from the non-lifelike movements of mineral particles to the much more life-like movements of pollen granules. The latter movements Brown himself, despite later disclaimers, had also first described in his notes as "in some cases as brisk as that of the analogous infusory animalcules."[104]

Bennett's Conversion to Spontaneous Generation

Charles Darwin's 1868 proposal of pangenesis as a mechanism to explain inheritance discussed ideas of histological "molecules" as models upon which he based his "gemmules." In this context, he discussed the Virchow *omnis cellula* view of cytology, in which he included Beale's "germinal matter." But Darwin also described "another school," which "maintains that cells and tissues of all kinds may be formed, independently of pre-existing cells, from plastic lymph or blastema . . . As I have not especially attended to histology, it would be presumptuous of me to express an opinion on the two opposed doctrines. But everyone appears to admit that the body consists of a multitude of 'organic units,' each of which possesses its own proper attributes, and is to a certain extent independent of all others."[105]

Darwin's tone gave the impression that the cytoblastema school was by this time a minority; however, he seems to have still felt that it was a school whose views deserved respect. Thus, when a new paper by Bennett came out six weeks later, arguing that new experiments and his own molecular theory pointed to heterogenesis,[106] the cytoblastema context was by no means considered discredited. And Bennett was only one among many who were arguing during 1868 and 1869 that Darwin's evolutionary theory required spontaneous generation as an underpinning, to avoid invoking non-natural causes for the origin of the very first organisms.[107]

In the natural theology–based scientific context of Britain, spontaneous generation theories were always on less firm ground than on the Continent. Thus, connecting the theory of heterogenesis to an already widely established discourse in medicine, histology, and microscopy gave such theories much greater appeal and respectability.[108] Bennett pointed out that no histologist had been included on the Commission of the French Academy that had been so biased against Pouchet, and that

Pouchet and his colleagues had thus been very proper in refusing to appear before a body whose investigation was "altogether one-sided and of no scientific value."[109] Even spontaneous generation opponent James Samuelson, editor of the new *Quarterly Journal of Science,* said that the French Academy's commission had behaved in such a way that its conclusions were not worthy of attention.[110]

Bennett further cited as a precedent for belief in spontaneous generation the 1839 article that caught Owen's notice, written by Allen Thomson, Bennett's predecessor in the Chair of the Institutes of Medicine at Edinburgh University.[111] Though he was one of the anti-radical Edinburgh set that Jacyna has termed "Philosophic Whigs,"[112] Thomson had in one of his more radical moments written that the balance of laboratory evidence (as of 1839) seemed to be "in favor of the occasional occurrence of spontaneous generation."[113]

In addition, Bennett's argument for heterogenesis appeared publicly at a moment of great energy in theoretical debates about infectious disease, following the recent cholera and cattle plague epidemics. Bennett's molecular theory of the origin of bacteria was surrounded in this context by many "particulate" theories of disease such as Farr's theory of "zymads," Béchamp's "microzymas," Beale's "germinal matter" (soon to be renamed "bioplasts"), and Burdon Sanderson's "microzymes."[114] Because many of these particles were descendants of Liebigite fermentation-style disease theory, only just beginning to be seriously challenged by Pasteurian germ theory,[115] Bennett's anti-Pasteurian position and his endorsement of heterogenesis could be viewed in British biomedical circles as fitting in with a familiar, long-standing discourse. Gilbert Child at Oxford was coming to identical conclusions at the same time.[116] Both men's reputations were important in overcoming spontaneous generation's disreputable history in Britain since the Crosse episode of 1837 and the publication of *Vestiges of the Natural History of Creation* in 1844. The momentum was further augmented when Richard Owen, who had been tentatively supporting Pouchet and heterogenesis, fully endorsed Bennett and Child's views in volume 3 of his influential *Anatomy of the Vertebrates* in November 1868.

Further Response to Owen's Conversion

In many ways Owen's new position anticipated the "physicalism" soon afterwards summed up in Huxley's famous lecture, "On the Physical Basis of Life." Huxley was attempting to showcase protoplasm as a substance not separated by an unbridgeable gap from the nonliving world. Indeed, Huxley's declaration that *Bathybius* on the sea bottom represented a primitive, poorly organized mass of protoplasm was read by many in England, as well as by Ernst Haeckel, as a proof of spontaneous generation.[117] George C. Wallich, a British surgeon and early oceanographer, criticized *Bathybius* as a construction by Huxley, far beyond what the facts justified.[118] Haeckel, however, went even further than Huxley, to state explicitly about *Bathybius,* "we cannot ponder this highly remarkable fact without the deepest astonishment, and cannot help thinking of Oken's 'primitive slime' . . . this . . . notorious primitive slime, whose all-embracing importance was indeed already implicitly established by Max Schultze's protoplasm theory, seems by Huxley's discovery of *Bathybius* to have become completely true."[119] Friedrich Lange interpreted Huxley in the same way. In 1875, in his widely read *Geschichte des Materialismus,* Lange commented: "Even among the Monera at present known there is a species which probably even now always comes into existence by spontaneous generation. This is the wonderful *Bathybius haeckelii,* discovered and described by Huxley."[120] As of 1875, Lange considered spontaneous generation as yet neither proven nor disproven, but seemed sympathetic to Pouchet and noted, like Owen, that "on Pasteur's side are the Academy and the Ultramontanes. To controvert the possibility of spontaneous generation is a mark of conservatism."[121] Lange describes another influential German researcher, Fechner, whose 1873 theory was very much in the tradition of the "Brownian molecules" line of thought. This theory was that "organic molecules" differ from inorganic ones by their state of mobility, i.e., by constantly changing their positions relative to one another, which state of motion is kept up by "inner forces" of the molecule.[122]

Because of the close resemblances between Owen's ideas and those of Huxley's physicalist approach by the late 1860s, most commentators tended to lump the two together. The review of Owen's book in the *Lancet* is just one such example. This occurred despite strong personal antagonisms that made both men deny that their ideas were similar. The medical journals lauded Owen's position on spontaneous generation, but disagreed with his claim that Darwinian evolution was opposed to this doctrine and ought to be lumped in with the "defeated" Pasteur and Cuvier. Henry Lawson, histology instructor at St. Mary's Hospital Medical School and editor of several scientific and popular science journals, explicitly linked Darwin's pangenesis with heterogenesis and lauded both.[123] He reprinted Bennett's 1868 article in the January 1869 issue of his *Popular Science Review.* Immediately following it, Lawson placed a review of *On the Anatomy of Vertebrates* that criticized Owen for attacking Darwin on trivial semantic points, while in substance "he thus, to our minds, differs but very little from the disciples of Mr. Darwin."[124] Regarding Owen's open conversion to Pouchet, Lawson said:

> This will pain and surprise not a few of his "creationist" supporters considerably. Yet we think it is the one "saving clause" in the volume, the one redeeming feature of a work which, however comprehensive, is so full of objectionable features . . . We are certainly of opinion that on this one point of spontaneous generation Professor Owen has allowed his mind to arrive at an unbiased conclusion, and in this solitary instance we think he is in advance of his confreres in this country, with the single exception of Dr. Hughes Bennett of Edinburgh, whose able essay in our present number is in great measure a demonstration of the principle of heterogeny. Professor Owen speaks his mind openly and honestly on this question, and lends the weight of his authority to the side of heterodoxy. But it is heterodoxy which we do not think we go too far in asserting will soon be very generally accepted.[125]

Thomas R. R. Stebbing, another enthusiastic young Darwinist, was soon to follow.[126] And Darwin's long-time friend and physician, Henry Holland, in an essay written around this time but only published posthumously in 1875, leaned toward Pasteur, but cautioned that Pouchet's view "cannot be put aside without further enquiry."[127]

Opponents of physicalism and spontaneous generation also saw a link between Huxley's and Owen's perspectives. Lionel Beale, for instance, in early 1869 attacked both men on their physicalism, and Owen especially on his analogy between an amoeba and a magnet. Beale, it should be noted, was a more extreme vitalist than many of his contemporaries.[128] He did, however, share a fear with Oxbridge conformists that the new "physical basis" and spontaneous generation arguments were headed down a more materialist road than even the one trodden by Darwin. While Beale's attack on Huxley has been examined at some length by Geison,[129] his simultaneous related attack on Owen has not, and thus bears examination.[130]

About the physical theory of life, Beale stated that if "even one unprejudiced person accustomed to weigh scientific evidence" was as yet unconvinced of the doctrine, then one must still consider unproven "the dogma that life is but a mode of ordinary force, and that the living thing differs from the nonliving thing not in quality, or essence, or kind, but merely in degree."[131] Among Owen's list of implications of physicalism, Beale focused on the analogy between an amoeba and a magnet as a particularly vulnerable target, asking:

> Is such reasoning as this likely to have weight with anyone who has seen an amoeba moving? . . . If the magnet moved itself from place to place; if it divided and multiplied; if every part of it were capable of moving in every direction; if it were able to select salts of iron, and then decomposed these and appropriated the iron to itself, so that, from a very little magnet, it grew into a big one, there would still be no real analogy between it and an amoeba; because you can magnetize and unmagnetize the steel as many times as you like, but you cannot revitalize an amoeba once defunct. If you were to take a quantity of dead matter of defunct amoebae and place it near a living amoeba, it would not be reanimated. The living amoeba might take up, bit by bit, the products of disintegration, and thus increase; but this is a very different thing from vitalizing a *mass of organic matter* as a mass of steel may be magnetized. Dead amoebal matter cannot be induced to *live* under circumstances at all parallel with those under which the "defunct" steel can be remagnetized. We are therefore compelled to conclude that the amoebal phenomena are different in their very nature from the magnetic phenomena.[132]

A few weeks later in the same periodical, Owen replied to Beale's attack. "In reference to the remarks by Professor Beale," he opened, "permit me to observe that there are organisms (Vibrio, Rotifer, Macrobiotus, &c.) which we can devitalize and revitalize . . . many times." He then expands from this to attack Beale's vitalism directly:

> As the dried animalcule manifests no phenomenon suggesting any idea contributing to form the complex one of "life" in my mind, I regard it to be as completely lifeless as is the drowned man whose breath and heat have gone, and whose blood has ceased to circulate. In neither dead body, however, is there rest: the constituent force-centres . . . are at work: a stagnant force-centre is to my mind inconceivable—a contradiction in terms. The change of work consequent on dying or

drowning forthwith begins to alter relations or "composition" and, in time, to a degree adverse to resumption of the vital form of force.[133]

Owen replied, further, that Beale was wrong that a piece of steel can be magnetized and unmagnetized for an indefinite period of time, but that since steel too is subject to the decomposing agencies surrounding it, the difference between it and an amoeba is only one of degree, not of kind. He accused Beale of taking a bantering tone in his critique, which did not aid in getting at the facts of the matter. Owen suggested that if, as he supposed, Beale opposed his claim that spontaneous generation ("the vitalization of the primary granule or jelly-speck") was possible, that "the analogies supporting the alternative belief . . . would be more instructive than banter."[134]

When James H. Stirling, an Edinburgh physician and philosopher, attacked Huxley's physicalist stance in his polemical *As Regards Protoplasm,* he believed, as Beale did, that Huxley was implying spontaneous generation (which Stirling called the "theories of the molecularists") in no uncertain terms. Stirling believed Huxley's purpose was to promote Darwinian evolution and the inseparably connected doctrine of spontaneous generation. He also made clear that the Crosse experiments of 1837 were still a bugbear: "Nay, the miracle they refuse at the hands of Moses, they are quite ready to accept at the hands of Mr. Crosse: they are quite ready to believe it possible for *him* to grind *wet* maggots out of *dry* electricity!"[135] He went on, clearly with Owen in mind, to argue against a view of "those who *feign* matter to be the expression of innumerable centres—whence, what, or how, one knows not!—of force."[136] And, describing the larger project he suspected behind it all, Stirling said:

> The . . . modes of theorizing indicated [spontaneous generation, physicalism, and evolution], indeed, are not without a tendency to approach one another; and it is precisely their union that would secure a definitive triumph for the doctrine of materialism . . . [T]he theory of the molecularists would, for its part, remove all the difficulties that, for materialism, are involved in the necessity of an egg; it would place protoplasm as formed from molecules, undeniably at length on a merely chemical level; and this theory being sound, would fairly enable Mr. Darwin, supplemented by such a life-stuff, to account by natural means for everything like an idea or thought that appears in creation. The misfortune is, however, that we must believe the theory of the molecularists still to await the proof.[137]

Nor did the *supporters* of spontaneous generation think such a connection between doctrines misguided, as we shall see. Not only Lawson, but also the much more influential Henry Charlton Bastian and many others argued strongly for the necessary linkage between spontaneous generation and evolution.

Spontaneous generation arguments in Britain up through the 1860s were very tightly connected to ideas about evolution in the mind of much of the public and scientific community alike. This was a strong linkage not only in the years just after the appearance of *Vestiges of the Natural History of Creation,* but continued after the publication of Darwin's theory. The arguments of Owen, Wyman, and Child showed that the linkage was firmly rooted in scientific views of evolution and suggested that Darwin and Huxley were merely keeping such views to themselves. In addition, the influential teaching and writings of Bennett had developed a strong connection between spontaneous generation and the doctrine of histological molecules, so that many perceived Darwin's gemmules to be within that context. Darwin himself had alluded to this, suggesting that his own ideas on spontaneous generation were still ambivalent. Bennett's authority as a technically up-to-date microscopist was important in counterbalancing the use of Beale's reputation in microscopy in opposition to spontaneous generation.

3

Bastian as Rising Star

Child, Wyman, Owen, and Bennett all made a certain impression on behalf of spontaneous generation. In addition, Lawson's reviews of Owen in December 1868 and January 1869 had solidified in the public mind the link between spontaneous generation and evolution. Thus, the uproar caused by Huxley's *Fortnightly Review* article, "The Physical Basis of Life," becomes intelligible: the article was perceived by most to be tantamount to Huxley's full endorsement of that linkage, and to add physicalism to the pot.

Beginning in early 1869 and running serially through the year in the *British Medical Journal,* there commenced a series of articles that made the case for the linkage of physicalism, evolution, and spontaneous generation more and more persuasively, steadily winning over in private and even publicly people of scientific importance. The identity of the author of these articles was not widely known until April 1870, but his case was sufficiently strong to attract the attention of Huxley, Busk, Sharpey, and Frankland, and to provoke a countercampaign by Tyndall. And once this author went public, long after Grant, Pouchet, Wyman, Bennett, Child, and Owen had either faded from the scientific limelight or died, Henry Charlton Bastian was seen by all of the English-speaking scientific world to be the most talented, eloquent, and vociferous evolutionist ever to make the case for spontaneous generation.

Bastian's Background

Henry Charlton Bastian was born in Truro, Cornwall on 26 April 1837. His father was a merchant who died when the son was still quite young.

However, Bastian apprenticed himself to an uncle who was a prominent doctor in nearby Falmouth. As a young man Bastian met a retired London surgeon in Falmouth who stimulated his interest in natural history. This led Bastian to publish a complete "Flora of Cornwall" at the age of nineteen, and a collection of all the ferns of Great Britain a year later, which won a prize from the Royal Cornwall Polytechnic Society.[1] In swift succession, Bastian obtained his B.A. (1859), M.A. (1861), and M.B. (1863) degrees at University College London (UCL). While working toward his M.D., Bastian carried out an extensive study in his spare time of the Guineaworm ("taken from the extremities of a well-known surgeon from Bombay")[2] and of the entire group of Nematoid worms, parasitic and free. The latter project resulted in a monograph in which a hundred new species were described. Bastian became an early and enthusiastic convert to evolutionary thinking by reading both Charles Darwin and Herbert Spencer during his university education.

Early on in his studies, Bastian stood out as a man of great scientific potential. He studied anatomy at UCL under G. Viner Ellis, zoology under Robert Grant, and physiology under William Sharpey. From Grant, as much as from Darwin or Spencer, Bastian received a spirited exposure to evolutionary ideas, though of a Lamarckian kind. But Grant was, in

Henry Charlton Bastian, approximately 1877.

addition, a strong advocate of spontaneous generation. Bastian won gold medals (including Grant's) in botany, comparative anatomy, anatomy and physiology, pathological anatomy, and medical jurisprudence. Although there is no doubt that Grant was a significant influence and maintained contact with Bastian after the younger man had become an advocate of spontaneous generation, I do not find convincing the claim that Bastian was merely "one of Grant's students, . . . still marching defiantly along their own evolutionary path."[3] Here I will argue for a significantly new interpretation of Bastian's position: as a mainstream Darwinian evolutionist, indeed as one of the most promising scientific talents in the next generation being mentored and shepherded by Huxley and the X Club during the late 1860s.[4] This does not imply that Grant, or his spontaneous generation ideas, had little influence on Bastian. Rather, just as Desmond has shown so convincingly that Grant's early influence on Darwin set him thinking and investigating ideas about evolution and spontaneous generation while still a young man, Grant's influence seems to have been similarly formative for Bastian. But in neither case did his pupils grow up to repeat Grant's beliefs in lock-step fashion. Both Darwin and Bastian went on independently to wrangle with theories of evolution and spontaneous generation. Bastian proves the depth of Grant's influence as a teacher for forty-five years at UCL, even though that influence did not create a coterie of Grant disciples.[5]

By 1860 Bastian had been appointed assistant conservator of the UCL Museum of Morbid Anatomy under Sharpey's direction.[6] He held this post for three years, until his departure to take up a residency in neurology at the State Asylum for Criminal Lunatics at Broadmoor from October 1863 until the end of 1865, under Sir John Meyer.[7] His decision in 1863 to study insanity was "principally on account of his liking for cerebral physiology and philosophical subjects generally."[8] By early 1866, Bastian had returned to London, qualified for his M.D. at the University of London, become engaged to Miss Julia Orme, and become assistant physician and lecturer in pathology at St. Mary's Hospital Medical School. By 1866, his position also included curatorship of the St. Mary's Museum of Pathology.[9] While there Bastian continued investigations he had begun at Broadmoor on the specific gravity of the brain. In addition he "now took up the study of the diseases of the nervous system as a whole, rather than the section of it met with in asylums."[10] During his eighteen months at St. Mary's, Bastian came to know Ernest Hart, the dean, and Henry Lawson, the teacher of histology and later lecturer on physiology, both of whom later offered support for his publications in journals they edited.[11] Francis Sibson, F.R.S., of the St. Mary's staff, wrote of Bastian's credentials

> It was a great pleasure to all of us to welcome to St. Mary's so able a man as . . . Dr. Bastian. I was already aware, as his Examiner in Medicine at the University of London, how sound and thorough a fund of Medical Knowledge he displayed. But his papers have given me the conviction that in Bastian our profession has a true engineer and worker, who by his concentrated research has already advanced science, and will do much to develop the science of Pathology. I count his accession to St. Mary's as the most important addition that our staff has yet acquired, and I feel that it will add to the growing prosperity of the school.[12]

Bastian's reputation, as indicated, was spreading among broader scientific circles, not just medical circles. On 15 June 1865, for instance, his memoir on the Nematoids was read to the Royal Society. It was proposed for inclusion in the *Philosophical Transactions,* and was thus assigned to two reviewers, T. H. Huxley and George Busk. Both Huxley and Busk were enthusiastic about the scientific talent of the young Bastian. Both cited the evident commitment of time and care to original investigations, as well as Bastian's disciplined systematic approach. They were also happy to see his commitment to evolutionary thinking, though varying in the degree to which they advised caution on speculative claims. Huxley's report stated that "Bastian's memoir . . . contains the results of a great deal of careful and troublesome original investigation, combined with an elaborate and conscientious survey of what has been done by other workers in this same field and I recommend its publication in the Philosophical Transactions." However, he cautioned that Bastian "would do well to revise his concluding remarks about the classification of the Nematoids as they contain one or two inaccuracies in matters of fact; and as their deduction that 'the Nematoids are simply aberrant Echinoderms' is one which in my judgment cannot possibly be sustained."[13]

Busk, likewise, concluded of the paper that "I considered it worthy of publication, from the circumstance firstly that the subject is one to which little attention has hitherto been paid . . . I stated also that Mr.

Bastian had communicated a large laborious systematic paper on the subject . . . to the Linnean Society, which had been published in its Transactions, and that this paper had shown that the author had successfully devoted much time and attention to the study of that class of animals." But about Bastian's speculative conclusions, Busk was less critical, saying: "although he goes further in his speculations with respect to their relations more especially with the Echinodermata than I would myself . . . I see no reason in this why he should not be allowed to express his own views in the way he has done."[14]

With two enthusiastic recommendations, the paper, including the speculations on affinity to the echinoderms, was voted for acceptance into the *Philosophical Transactions* on 26 April 1866.[15] On the strength of his research publications, Bastian was proposed for Royal Society membership by Busk on 7 February 1867. The petition for membership was signed by Busk, Darwin, Lubbock, William Carpenter, William Flower, Lionel Beale, and James Glaishier, president of the Royal Microscopical Society[16] among others, as well as several medical men, including Sibson, John Marshall, and neurologist J. Lockhart Clarke. After the collection of sufficient signatures, Bastian was elected F.R.S. on 4 June 1868.[17] With such widespread support among the Darwinians, it seems clear that they viewed Bastian as an evolutionist of considerable potential. His memoir was also extracted in the journal that this "Young Guard" currently looked to as their organ of expression, *The Reader.* Indeed, Roy MacLeod lists Bastian as a solid member of that young guard,[18] and Gordon Holmes describes him as "one of the group that included Darwin, Russel Wallace, Huxley, and others which finally established the theory of evolution," as well as "a friend and later the literary trustee of . . . Herbert Spencer."[19]

In late 1867 Bastian won appointment as professor of pathological anatomy in the medical faculty back at his alma mater, University College, as well as becoming assistant physician at University College Hospital.[20] Within a few months he had caught the eye of John Russell Reynolds, professor of medicine at University College, and become his protégé, both at UCL and as assistant physician at the National Hospital for the Paralyzed and Epileptic at Queen Square.[21] Reynolds commissioned Bastian to write a number of sections on pathology and morbid anatomy of various brain and spinal cord disorders for his new text, *A System of Medicine.*[22] Another young physician who joined the staff at

Queen Square shortly before, Thomas Buzzard, reminisced with William Gowers about their colleague Bastian's rapidly rising stature at this time, saying that Gowers and Bastian brought "into the hospital an atmosphere of scientific precision and method" that contrasted sharply with the "decidedly commercial tone" with which the hospital had been run up to that point.[23]

Bastian's research continued in the area of pathological neurology, including aphasia (speech loss due to neurological damage).[24] He took part in the debate with Alexander Bain over the "muscular sense"[25] and was a leading advocate of the doctrine of psychophysical parallelism in England.[26] Bastian's interest now also moved into the realm of more basic physiology, alongside Michael Foster under the eye of Sharpey in the lab at University College. He took a great interest in the phenomenon of the movement of white blood corpuscles out of capillaries into the surrounding body fluid, later called diapedesis.[27]

Bastian had also published technical articles on the details of staining procedures he had developed.[28] By July of 1868, Arthur Durham was lauding Bastian's skill as a microscopist, in his presidential address to the newly formed Quekett Microscopical Club, while in the same breath touting microscopy as guaranteed to develop "the Moral Qualities," most of all patience: "We cannot all of us be Queketts, Carpenters, or Beales, but every one of us can do something—we can advance ourselves, and we can help those around us. And if what I have said be true, everyone who works well with his microscope during such opportunities as he may have, cannot fail to become in more senses than one 'a wiser and a better man.'"[29]

Thus, as both distinguished physician and Victorian man of science, Bastian's reputation was widely spread by the relatively young age of thirty-one.

Bastian Enters the Spontaneous Generation Debate

Bastian's established place among the young guard can be gauged, among other things, by the fact that Alexander Macmillan in Britain and Edward L. Youmans in America both eagerly sought to publish his works. These men were noted from the 1860s onward as seeking to advance the cause of evolutionary science, as well as of public education in science; essentially they were in full agreement with the "young guard" and saw Tyndall and Huxley as ideal models of how to put this agenda into practice.[30] Youmans had cordial relations with Bastian, as well as with the older Darwinians, and was always scouting for young talented authors whose works he could publish to further the cause of scientific naturalism. He edited the *North American Review* and later began *Popular Science Monthly,* using both as venues for this cause.

It was Youmans, for instance, who spotted young Spencer enthusiast John Fiske at Harvard and recruited him,[31] and who also published a critique of Spencer by Francis E. Abbot in October 1868.[32] The latter specifically challenged Spencer to be consistent with the "first principles" of evolutionary naturalism and to admit that spontaneous generation was necessarily implicit, just as much as evolution was. Spencer forcefully denied the possibility of rapid spontaneous generation, and insisted that "the facts and arguments [of his denial] had the unqualified endorsement of Huxley, Tyndall and Frankland."[33] Youmans was interested in foregrounding the potential inconsistency to which Abbot pointed, and he seemed to feel that the question was open still for experimental resolution. Thus, when Spencer wrote him in response to Abbot's essay asking for publication of a rebuttal,[34] Youmans replied, "Is it, in fact, needful for you to commit yourself to either side of the question at present contested?"[35] Nor was Youmans alone among those sympathetic to evolution in drawing this conclusion. As I have already shown, a great many others were making similar arguments. Spencer admitted as much:

> the same . . . objections have already been made in England—the one by Dr. Child of Oxford, in his *Essays on Physiological Subjects,* and the other by a writer in the *Westminster Review* for July, 1865 . . . Indeed, the fact that Dr. Child, whose criticism is a sympathetic one, puts the same construction on [Spencer's hypothesized "physiological units"], proves that your reviewer has but drawn what seems to be a necessary inference. Nevertheless, the inference [of the necessity of spontaneous generation] is one which I did not intend to be drawn.[36]

By mid-1869, the debate was in full swing. Despite Spencer's explicit rejection of spontaneous generation, which remained unpublished until 1870, the public perception of Huxley's position (after his discovery of *Bathybius* in August 1868 and his paper, "The Physical Basis of Life," appeared in February 1869) was that he supported the idea of spontaneous generation. Supporters of Huxley concurred. Thus, it was not surprising when an anonymous series (written by Bastian) in the *British Medical Journal* explicitly pressed this point. The series included "The Doctrine of the Correlation of the Physical and Vital Forces," "Vital Functions and Vital Structures," and the seven-part "The Origin of Life." These all appeared in a revised form in 1872 as part 1 of volume 1 of Bastian's best-known book, *The Beginnings of Life.* The articles argued for applying the physicalist doctrines of Huxley and Tyndall with the principle of continuity in nature. This strategy implied that the boundary between living and nonliving was just as likely to be crossed by natural processes as was the boundary between electricity and magnetism or heat.[37]

Alison Adam has discussed the path of Bastian's experimental and clinical work that led him toward spontaneous generation.[38] Here I wish to emphasize the broader theoretical concerns of evolutionary naturalism, plainly displayed in the *Bathybius* debate and in his articles in the *British Medical Journal*, which seem at least as important as Bastian's microscopic observations in steering him toward spontaneous generation. John Browning, the widely respected scientific instrument maker, agreed with the basic position of Bastian's *British Medical Journal* series, saying, "There is no boundary line between organic and inorganic substances . . . Reasoning by analogy I believe we shall before long find it an equally difficult task to draw a distinction between the lowest forms of living matter and dead matter."[39]

As described by a status report on the debate of 23 July 1869 in the *Journal of the Quekett Microscopical Club,* Bastian's first few installments in the *British Medical Journal* were found persuasive and widely accessible.[40] Club president Arthur Durham seemed fully willing to grant serious consideration to Bastian's arguments that the possibility of spontaneous generation was fully in line with tenets such as evolution and the "correlation of forces" (related to the conservation of energy). He may still have been unaware that Bastian was the author of that anonymous article. Thomas Clifford Allbutt, a colleague of Bastian who was also rising rapidly in medical circles,[41] was of a similar opinion, especially over the principle of continuity implying spontaneous generation as likely:

> I had attributed the papers in the *BMJ* on life to you: I was very much pleased with them. I have always thought it was cool of the anti "spontaneous" people in the present state of science to throw the a priori unlikelihood upon their adversaries. Surely it is the continuity people

> who have the a priori likelihood. I see Lister in his opening lecture lays great stress upon Pasteur's experiments. The expression "spontaneous generation" too is dyslogistic, and fastens on an absurdity they don't hold on evolutionists. I am very glad you are publishing a separate volume on the matter. As science advances so far as to change prevalent systems of thought, a priori probabilities must be corrected too.[42]

A similar assessment was given in an editorial by Darwinian sympathizer Henry Lawson in the 28 April 1869 issue of the weekly *Scientific Opinion,* in which he reflects

> It seems to us a little strange that many among the fiercest opponents of spontaneous generation are yet most implicit believers in the law of natural selection, and, indeed, in the general principles of evolution. Why this is so we cannot comprehend . . . Surely it is more in accordance with the general scheme of evolution to admit that organized matter is capable when broken up of reproducing itself, so to speak, than to assume that it must have been separately and specially created . . . On merely *à priori* grounds, we cannot see how the Darwinian disciples can reject heterogeny.[43]

As discussed in the previous chapter, by this time Beale and Stirling had also launched their aggressive counterattacks against Huxley's physicalism, and against the implied doctrine of spontaneous generation. Bastian, hoping to moderate, attempted to get a less dogmatic response from Stirling by writing to him privately. Though Stirling was effusive in his reply about all the accolades he'd heard about Bastian from Edinburgh professors Masson and Grainger Stewart, the letter indicated that the Scot was more rigid than ever in not giving Huxley credit for anything but confusion in his logic.[44] Bastian must have despaired of any compromise with Stirling, for, in a scathing review of Stirling's pamphlet early in 1870, he defended Huxley, the "molecularists" (Owen and Bennett), and "physicalists" in a manner that showed he believed himself, Owen, and Huxley were all in agreement over spontaneous generation because of the connection with evolution pointed to by Stirling. He went further to state that Stirling

> evidently believes that the doctrines of the "molecularists" concerning the new evolution of living things will have long to "await the proof"; but we . . . firmly believe the time to be not far distant when this will be as much an accredited dictum of science as are the doctrines of the

> Correlation of the Physical Forces, and of the Correlation of the Vital and Physical Forces which have been its necessary predecessors. We would ask [Stirling and his ilk], at least, to speculate upon the possibility of this. Let them learn in the meantime how they may best readjust their doctrines, so that when the time comes . . . there may be no sudden bewilderment, no feeling as if the very ground were being swept from underneath their feet.[45]

It must have been surprising then for Bastian when Huxley, at first helpful and sympathetic to him, soon began to distance himself more and more from Bastian, as the young pathologist made his claims about spontaneous generation more public and more forcefully. This began with Bastian's first paper on his experimental work, published in *Nature* on 30 June 1870, which I will describe in the next chapter.

Jeffries Wyman, a highly respected evolutionist at Harvard University, agreed with the criticisms of Huxley's and Spencer's inconsistency.[46] When he learned of Bastian's experiments on a visit to London in 1870,[47] Wyman read Bastian's papers on the subject and wrote to encourage the younger man to pick up where he had left off in attempts to solve the spontaneous generation problem experimentally:

> since reading your own results I have thought it far wiser to leave to others the battle. A crucial experiment bearing upon the subject of spontaneous generation, as I have found from sad experience, is most difficult. My primary standpoint is this: if there ever was a time when organic life did *not* exist on the surface of the earth, the transition to the period when it *did* exist was through spontaneous generation. If the question is approached from a scientific point of view I see no other alternative. The experimental proof may be slowly completed, but I believe the cumulative evidence in favor of it is becoming day by day stronger.[48]

Other Darwinians continued to argue that spontaneous generation was necessary along with evolution for any naturalistic science to be consistent and for continuity in nature. Rev. Thomas Stebbing, for instance, wrote: "It does not . . . seem incredible that living organisms, simpler perhaps than any yet detected by the microscope, should be . . . produced without generation by the mere combining of inorganic materials. This is the hypothesis of Spontaneous Generation, so called, or *abiogenesis,* unproved and extremely difficult of proof, but precisely filling that gap in the order and continuity of nature which is so puzzling without it . . . The theory [of evolution] does not deny the perpetuation throughout vast ages of extremely simple organisms."[49] Stebbing argued for a Lamarckian "escalator" view of evolution: as simple organisms evolved into more complex ones, they would continuously be replaced by spontaneous generation.

Wallace and Darwin Discuss Bastian

Alfred Russel Wallace was so persuaded by Bastian's evidence and powerfully argued case that he glowingly reviewed Bastian's *Beginnings of Life* when it appeared in 1872. He especially noted Bastian's argument that continuous creation of new microbial life and rapid transformations by heterogenesis could greatly speed up the rate at which evolutionary change occurs. This, he suggested, could provide an answer to the stultifying challenge of William Thomson's claim that earth had not existed in a cooled state long enough to allow the amount of time required by Darwin in *On the Origin of Species* for evolution to have occurred.[50] Wallace continued, "It is very strongly argued by Dr. Bastian that the conception of an origin of living organisms at a single remote epoch in past time, and the lineal descent of all existing organisms from those primal forms, is one quite opposed to the uniformitarian and the evolutional philosophy."[51]

Charles Darwin himself, after his careful reading of *The Beginnings of Life* at Wallace's urging, concluded that Bastian "seems to me an extremely able man, as indeed, I thought when I read his first essay." He went on to say:

> His general argument in favour of Archebiosis is wonderfully strong, though I cannot think much of some few of his arguments. The result is that I am bewildered and astonished by his statements, but am not convinced, though on the whole, it seems to me probable that Archebiosis is true. I am not convinced partly I think owing to the deductive cast of much of his reasoning; and I know not why, but I never feel convinced by deduction, even in the case of H. Spencer's writings. If Dr. B's book had been turned upside down, and he had begun with the various cases of heterogenesis, and then gone on to organic and afterwards to saline solutions, and had then given his general arguments, I should have been, I believe, much more influenced. I suspect however that my chief difficulty is the effect of old convictions being stereotyped on my

> brain . . . Perhaps the mere reiteration of the statements given by Dr. B. by other men whose judgment I respect and who have worked long on the lower organisms would suffice to convince me. Here is a fine confession of intellectual weakness; but what an inexplicable frame of mind is that of belief! . . . I should like to live to see archebiosis proved true, for it would be a discovery of transcendent importance . . . If ever proved, Dr. B. will have taken a prominent part in the work. How grand is the onward rush of science; it is enough to console us for the many errors we have committed and for our efforts being overlaid and forgotten in the mass of new facts and new views which are daily turning up.[52]

Wallace replied, "I quite understand your frame of mind & I think it a natural & proper one. You had hard work to hammer your views into people's heads at first, & if Bastian's theory is true he will have still harder work, because the facts he appeals to are themselves so difficult to establish."[53]

Edward Youmans was reinforced in his high opinion of Bastian by Wallace's strong support. He sought from Macmillan the rights for an American edition of Bastian's forthcoming book, *The Beginnings of Life*.[54] He later ran a review of *The Beginnings of Life* in the November 1872 *Popular Science Monthly* that echoed Wallace's enthusiasm and cited Wallace's opinion as proof that spontaneous generation was fully compatible with evolution.[55] Youmans also continued to cordially encourage Bastian in his efforts in the spring of 1874, printing Bastian's argument[56] in installments in the *Popular Science Monthly* and saying: "I am glad to see that you keep up the scientific fight so vigorously: your first paper was read with a good deal of interest here by many whom I know and has been recognized as making out a powerful case, and I see by glancing over the second paper that the argument is strengthened. I think you give plenty of business to our friends Spencer and Huxley."[57]

As late as 1875, Youmans still felt strongly enough about Bastian's scientific credentials to make him "Scientist of the Month" in the *Popular Science Monthly*,[58] to sign him on, along with Tyndall and Huxley, to write a book for the new International Scientific Series, and to pay him an advance of £100.[59] And when Tyndall first began to attack Bastian's scientific reputation in public, Youmans seems to have urged him to behave in a more respectful fashion toward Bastian, regardless of their differences.[60]

It should be noted that Wallace later came to see Bastian's position on spontaneous generation as unattractive, though not for the same reasons as Huxley and Tyndall. Especially after Bastian had become widely known as a proponent of psychophysical parallelism, again one of the essential principles later associated with "scientific naturalism," Wallace found Bastian's overall position completely incompatible with, indeed hostile to, the spiritualism that had become so central to his own worldview. Thus, by the early twentieth century, these two heirs of evolutionary theory demonstrated just how far different "Darwinisms" could diverge. In this light, Gordon Holmes noted that "Wallace, who was an ardent believer in spiritualism, stated when an old man that in his opinion Bastian possessed one of the most acute intellects he had met, and spoke of him as unrivalled in argument and debate, but added it was incomprehensible how such a man could believe in abiogenesis. When a few days later Wallace's greetings were conveyed to Bastian he spoke of Wallace as a remarkable scientific observer, and commented, 'How can such a man believe in spiritualism?'"[61]

Further Support for Bastian

Another independent source of support for Bastian's ideas appeared in April 1872, in an article by James Murie, sometime librarian of the Linnean Society. Murie's paper reported finding microorganisms within the chest cavity of living birds immediately upon opening it, and he concluded "a fresh interest accrues in those cases where epiphytes arise within closed cavities of the animal body . . . Do such phenomena bear upon those doctrines of 'spontaneous generation'?"[62] Murie was acquainted with Bastian and, incidentally, asked him at about this time to write a supporting letter for Murie's candidacy for the professorship of general and comparative physiology at the Royal Veterinary College. Bastian complied.[63]

A useful indicator of the prevalence of positive assessments of Bastian is the comment in the *Annual Register* at year's end in 1872:

> The subject which stands out pre-eminently this year as riveting the attention of men of science, and producing wonder in the minds of those who have but to take the results of investigation and analysis as they are propounded by the skilled experimentalist, is the spontaneous generation doctrine advocated, and it is said all but established, by Dr. Bas-

> tian . . . Whether or not Dr. Bastian's statements of fact are all capable of verification, it seems to be generally admitted that a great stride has been made in biological science by his investigations, and that a further elucidation has been attained of *that unity and continuity of Nature's laws which is so marked a result of modern scientific research.*[64]

Thus it is clear that many evolutionists and supporters of that cause saw Bastian's version of scientific naturalism *with* spontaneous generation as an equally valid competitor to Spencer and Tyndall's version without it, perhaps as having an even better claim to be the version most compatible with the doctrine of continuity. Further, Bastian's scientific credentials were widely viewed as more than sufficiently strong to entitle him to challenge Huxley, Tyndall, and Spencer as peers.

Bastian's reputation continued to grow among medical circles as well, based on his clinical work as well as his scientific work. In 1870 he was elected to membership in the Royal College of Physicians.[65] William Osler, a young medical student at this time, reported that

> in the summer of 1872 after a short *Rundreise,* Dublin, Glasgow, and Edinburgh, I settled at the Physiological Laboratory, University College, with Professor Burdon Sanderson, where I spent about fifteen months working at histology and physiology. At the hospital across the way I saw in full swing the admirable English system, with the ward work done by the student himself the essential feature. I was not a regular student of the hospital, but through the kind introduction of Dr. Burdon Sanderson and of Dr. Charlton Bastian, an old family friend, I had many opportunities of seeing Jenner and Wilson Fox, and my notebooks contain many precepts of these model clinicians. From Ringer, Bastian and Tilbury Fox, I learned too, how attractive out-patient teaching could be made.[66]

Bastian's highly respected colleague Burdon Sanderson, formerly a student of Bennett,[67] wrote and spoke on behalf of Bastian's great skill as an experimentalist. As described in chapter 1, Burdon Sanderson's claim to have replicated some of Bastian's experiments in January 1873 was one of the most powerful testimonies of any colleague on his behalf. This continued to be so, even though Burdon Sanderson insisted he did not necessarily agree with Bastian's interpretation of the results that both men saw. The young Osler, while working in Burdon Sanderson's lab, carried out some research he thought confirmed Bastian's claims on heterogenesis. With Burdon Sanderson's support the paper was read to the Royal Society on 18 June 1874.[68]

Jabez Hogg, a noted physician and author of a widely used text on microscopy, was a supporter of Bastian's theories of disease,[69] as was Timothy Lewis, who specialized in research on cholera.[70] William Sharpey continued to speak highly of Bastian's work, for example to the famous sanitary reformer Edwin Chadwick.[71] Bastian was also courted to write numerous articles on neurological and pathological subjects, including bacteria and the germ theory of disease, for Richard Quain's new *Dictionary of Medicine*.[72]

Bastian was elected Fellow of the Royal Society at an unusually young age, and his very first paper to that body was accepted with acclaim for publication in its prestigious *Philosophical Transactions*. Thus, he had looked to the Society from the very beginning as an important forum for publicizing his views. He had told the noted publisher of the scientific young guard, Alexander Macmillan, that presenting a paper at the Royal Society would be his first major public move in promoting his views on spontaneous generation.[73] The paper was not actually presented when Bastian realized that, "owing to the accumulation of many papers and other causes, no evening could be allotted on which it might be read and discussed."[74] However, feeling that there was need to get an in-depth statement of his position and experiments before a scientific audience without undue delay, especially after his public confrontation with Tyndall in the *Times* in April 1870, Bastian opted to submit the paper to *Nature*. Norman Lockyer, the editor, was on friendly terms with Bastian, and Macmillan, its publisher, had been planning imminently to publish Bastian's book-length treatment of the subject for more than six months, advertising it as forthcoming in the pages of *Nature* since January.[75] Thus it was agreed that Bastian's long paper would appear *in extenso* in three installments in the 30 June and 7 and 14 July issues.

Bastian did not give up on the Royal Society, however. As the spontaneous generation controversy became public and outspoken, he applied for a research grant in the fall of 1870, which Sharpey as secretary for the biological sciences passed on to the Society Council for him. Bastian was also able to schedule the reading of several of his experimental papers at the Thursday evening meetings of that prestigious body. These included: "On Some Heterogenetic Modes of Origin of Flagellated Monads, Fungus Germs, and Ciliated Infusoria," which was received on 15 February

1872 and read on 21 March;[76] "Note on the Origin of Bacteria, and on their Relation to the Process of Putrefaction," which was received on 20 November 1872 and read on 30 January 1873;[77] "On the Temperature at Which *Bacteria, Vibriones,* and their Supposed Germs are Killed when Exposed to Heat in a Moist State, and on the Causes of Putrefaction and Fermentation," read on 20 March 1873; and "Further Observations" on the same subject, read on 15 May 1873. All of these papers were accepted very quickly, by vote of the Society, for publication in the *Proceedings.*[78]

Bastian's claims did not go uncontested, however. By 19 June 1873, Henry Acland, the Regius Professor of Medicine at Oxford University communicated a paper by two of his students, C. C. Pode and E. Ray Lankester (also a young protégé of Huxley), challenging Bastian's experiments on numerous technical points. This will be taken up in the next chapter.

Within a year of entering the British debate publicly, Bastian had become the main figure supporting spontaneous generation, with Owen, Child, Wyman, and Bennett having withdrawn to much more peripheral roles by the end of 1870. Bastian's stature in the medical and laboratory science communities, and in the evolutionary "young guard" gave the doctrine a level of publicity previously unheard of in Britain. As the next chapter will show, the same year saw the consolidation of a new vocal opposition as well. Interestingly, this was not from the "conservative" scientists such as Beale or Stirling, but centered on Huxley and Tyndall, who launched a major campaign to publicly separate Darwinian evolution from any necessary connection with present-day spontaneous generation. Because of his high profile, Bastian was the lightning rod upon which most of this reaction was discharged. Thus, though there was substantial continuity in the issues under discussion from 1860 through the late 1870s, especially the connection between spontaneous generation and evolution, the next phase of the British debate was distinguished by a fairly sharp shift in personnel as well as by sharply increased public visibility, beginning with the entry of Bastian.

4

Initial Confrontation with the X Club: 1870–1873

As the spontaneous generation debate heated up in Britain in late 1869 and early 1870, the argument linking evolution with spontaneous generation was one of the most influential in the armory of advocates such as Ernst Haeckel and Henry Charlton Bastian. Tying spontaneous generation to the rising fortunes of Darwinism made theoretical sense to many evolutionarily inclined British men of science, and thus was an effective strategic move. Only when the most widely recognized Darwinians, those in the so-called X Club, came out publicly against this (as they considered it) unholy alliance, did the fortunes of spontaneous generation begin to slow.

Huxley's Tightrope Act

The X Club's clash with Bastian began with John Tyndall's January 1870 lecture at the Royal Institution, "Dust and Disease." Since Tyndall the physicist was seen by many as an interloper into biology and especially medicine, where he had no expertise, much more important was T. H. Huxley's September 1870 BAAS presidential address in Liverpool and another paper he delivered at that meeting, both of which unequivocally opposed any alliance between *present-day* spontaneous generation and evolutionism. Huxley went beyond his earlier caution about Bastian's speculation being insufficiently grounded (as in the Nematoid paper), and now specifically criticized Bastian's experimental work as well. The result was a power struggle between pro– and anti–spontaneous generation factions among Darwinians that waxed hot through the 1870s. Bastian emerged as the leader of the "pro" faction in opposition to the X Club largely because of his prolific experimentation, his rhetorical abilities, and his influence in the medical community. There was widespread resentment among medical men toward the pronouncements about disease from nonpractitioners such as Tyndall and Huxley, who had no clinical experience of disease.[1]

To recap briefly, after his discovery of *Bathybius* in August 1868 and his paper in February 1869, "The Physical Basis of Life," the public perception was that Huxley supported the idea of spontaneous generation. James Moore has summed up Huxley's difficult public relations dance very well. Ernst Haeckel, the most vociferous advocate of Darwin's ideas in Germany, immediately began to declare that *Bathybius* proved that spontaneous generation was a necessary correlate to evolution.[2] Back in England, meanwhile, through 1868, George H. Lewes was still writing in the radical *Fortnightly Review* defending Haeckel's spontaneous generation theory and claiming that "Mr. Darwin has reason to be proud of his disciple."[3] We should also note that Stirling, Beale, and the medical journals, from 1869, kept up the public association of evolution, physicalism, and spontaneous generation, creating a climate toward which Huxley could try to appear cool, but by which he could not remain unaffected. In early 1871 the appearance of Darwin's *Descent of Man* fanned the flames of accusation that Darwin's project was indeed atheistic after all. Though he did not respond to Stirling until December 1871, and even then only with a few dismissive lines,[4] Huxley's position defending Darwinism was made more difficult, so that more public distancing from the most fervent spontaneous generation advocates was required.

Huxley's strategy is evident in his construction of the category "agnosticism" at about this time to describe his views, as well as in careful positioning in his September 1870 BAAS address at Liverpool, "Biogenesis and Abiogenesis."[5] There he left the spontaneous generation option hypothetically open, though preferably in the distant past only, as much as seemed logically necessary for a defense of Darwinism.[6] Meanwhile, in another session at the Liverpool meeting, he pointedly criticized Bastian for careless experimental technique and for failing to properly interpret Brownian movement as nonliving in character.[7] The thrust of Huxley's view of Brownian movement will be discussed in this chapter. Whatever his feelings about Bastian's experimental competence, however, the larger need to keep Darwinism respectable in a public arena, not associated with such names as Crosse's, was clearly an important force behind Huxley's maneuvering. Busk and Huxley had been able to disagree in their earlier assessment of how far Bastian had a right to go with his speculations, given the amount of effort he had invested. But the stakes over the spontaneous generation question were much higher, and this required tightening the X Club ranks to allow no serious differences to show in public.

Even after Huxley's repudiation of his claims about *Bathybius*, George C. Wallich, an oceanographer and microscopist who had criticized *Bathybius* as a construction of Huxley's from the beginning, continued to urge the need for Darwinism to expressly dissociate itself from Haeckel's spontaneous generation as late as 1882. In a letter to Darwin not long before the famous naturalist's death, Wallich said, "In a lecture I am about to give on the Threshold of Evolution, in which I dispute *in toto* Haeckel's statements concerning the Protista, I am anxious to show that a statement put forward by him and others—'that spontaneous generation is necessary to the completeness of evolution as a doctrine'—has nowhere received your caution."[8]

Another supporter of Darwin, who was trying to prevent conflict between evolutionary doctrine and religion, was Methodist minister and biologist William Dallinger. Dallinger belonged to the Christian Evidence Society, a group of "moderate and evangelical churchmen . . . who were not identified with either the ritualist or the rationalist extremes of the Church of England," and who sought defense of the faith against doubt and atheism by seeking evidence for "new affirmations of the certainty of Christianity."[9] The presence of such men as Dallinger among its ranks "helped to ensure that the Society could not be fairly charged with being hostile to new discoveries or to speculations concerning the origins of the universe."[10] Dallinger took an active role in looking for evidence of "life cycles" among the protozoa or "monads," since this could powerfully undermine spontaneous generation claims for microbes just as it had for parasitic worms shown to develop through many stages dissimilar in appearance to one another. At the height of the spontaneous generation controversy in Britain in the 1870s, he and a friend, John Drysdale of Liverpool, published a series of articles in the *Monthly Microscopical Journal* that very influentially made just that argument.[11] This work will be discussed in detail in chapter 5.

In addition, although an editorial in the new journal *Nature* suggested skepticism about the atmospheric germ theory in the British scientific community,[12] Darwinian X Club member John Tyndall had responded by committing himself very prominently in the *Times* (in the letter of 7 April 1870 that provoked a reply from Bastian) to atmospheric germ theory as advocated by Pasteur and Lister. Especially after the extreme position that Tyndall took in response to Bastian's sharp criticism, Huxley's attempt to find a posture on spontaneous generation that did not logically undermine Darwinian evolution was made more difficult. He wished to find a way that would not contradict outright and humiliate his X Club ally over the issue of Tyndall's germ theory of disease. For, while many argued that the germ theory from a medical point of view need not necessarily imply any definite position on the question of spontaneous generation,[13] Pasteur argued forcefully for a linkage of the two issues. If one accepted the germ theory, one rejected spontaneous generation—a linkage that both Tyndall and Bastian had accepted. Thus, their opinions on the origin of disease powerfully conditioned their views on spontaneous generation, with Tyndall strongly supporting Lister,[14] and Bastian advocating an explicitly Liebigian chemical fermentation theory of disease.

Yet Huxley and Busk had both been very enthusiastic about Bastian's work. They continued to take an interest even when he first claimed to find experimental proof of spontaneous generation, as did Frankland and William Sharpey. So how did the transition to declaring Bastian's work incompetent occur? To understand this, we need to look briefly at Huxley's manner of guiding the careers and *Bildung* of rising young evolutionary scientists.

Huxley's Attitude toward Young Men of Talent

Huxley viewed himself as mentor and guide to a number of young scientists interested in evolution in the late 1860s and early 1870s. These included Ernst Haeckel, Anton Dohrn, William Flower, Michael Foster, Alexander Kowalevsky, Nikolai Miklucho-Maclay, St. George Mivart, E. Ray Lankester, and the philosopher John Fiske, as well as Bastian. Huxley was attentive not only to their experimental and theoretical progress, but to the development of their characters and to acquainting them with his idea of the proper code of behavior for professional evolutionary scientists. An early exchange with Anton Dohrn illustrates this quite well. Dohrn, at twenty-eight, had been corresponding with Huxley for some time, and had developed a theory on arthropod evolution. He was so excited about his theory that he sent an article off to be published even though Huxley had cautioned him that the theory seemed shaky and advised waiting for further evidence and reflection before hasty publication. (Recall Huxley's very similar warning to Bastian in his 1865 reviewer's report.) By the time the article appeared in 1868, Dohrn had reconsidered and felt very sheepish at not having followed his mentor's guidance. Huxley responded

> As you know, I did not think you were on the right track with the arthropods, and I am not going to profess to be sorry that you have finally worked yourself to that conclusion. As to the unlucky publication in the Journal of Anatomy and Physiology you have read your Shakespeare and know what is meant by "eating a leek." Well, every fine man has to do that now and then, and I assure you that if eaten fairly and without grimaces, the devouring of that herb has a very wholesome cooling effect on the blood—particularly in teaching sanguine temperament. Seriously, you must not mind a check of this kind.[15]

This was a general principle for Huxley, not just an ingenious onetime literary trope. Huxley repeatedly used the "leek eating" expression to express that particular rule of scientific decorum to his young protégés. For instance, in 1875 when *Bathybius haeckelii* was found to be an artifactual chemical precipitate from seawater by the scientists aboard the Challenger expedition, and not the primitive protoplasmic form Huxley had declared it to be in 1868, Huxley wrote punningly to Michael Foster, "The 'Challenger' inclines to think that *Bathybius* is a mineral precipitate! in which case some enemy will probably say that it is a product of my precipitation . . . Old Ehrenberg suggested something of the kind to me, but I have not his letter here. I shall eat my leek handsomely, if any eating has to be done."[16] At the 1879 BAAS meeting at Sheffield, when the issue of Huxley's mistake was brought up again, he again gracefully ate the leek, according to his *Life and Letters*.[17] Foster showed that the lesson had been internalized when he wrote to "my Lord Mayor," "By the bye, you did that *Bathybius* business with the most beautiful grace—I wish you would sell me a little morsel of that trick."[18] Some at the Sheffield meeting did not agree that Huxley's owning up to his error was done so graciously; for instance the physiologist William Carpenter reported that the meeting "was not good. That Huxley seemed sore under mention of *Bathybius*."[19]

In another letter to Dohrn, Huxley maintained his earlier humorous and fatherly tone, but described a bit more explicitly his self-appointed role as critic of the younger Darwinians:

> What between Kowalevsky and his ascidians, Miklucho-Maclay and his fish brains, and you and your arthropods—I am becoming schwindlesuchtig, and spend my time mainly in that pious ejaculation "Donner und Blitz," in which you know I seek relief. Then there is our Bastian, who is making living things by the following combination . . . Now I think that the best service I can render to all you enterprising young men is to turn devil's advocate, and do my best to pick holes in your work . . . My friend Herbert Spencer will be glad to learn that you appreciate his book. I have been his devil's advocate for a number of years, and there is no telling how many brilliant ejaculations I have been the means of choking in an embryonic state.[20]

In evaluating Huxley's tactics as mentor to this stable of young talent, it is important to note two crucial events of 1869. One was the increasing furor over spontaneous generation spurred by his paper, "The Physical Basis of Life." Another was the defection, on 15 June 1869, of St. George Mivart, one of the brightest young Darwinians Huxley was grooming. Mivart during the next two years published some of the most technically well-informed, and therefore damaging, critiques of Darwinism of any yet produced. And it was no surprise, since he had for years been studying at Huxley's elbow, acquiring the technical expertise that made his treason so publicly damaging to the cause of Darwinism. This surely made Huxley exquisitely sensitive during these months to any behavioral impropriety among the younger ranks, impropriety that might hint of another damaging defection to come. It is in the context of Huxley's larger role as self-appointed policeman of the younger ranks, and of his heightened sense of having completely overlooked a defection-in-the-making during the last months of 1869 and the opening months of 1870, that his encounters with Bastian at that time must be viewed.

Bastian had corresponded with Tyndall about his experiments as early as January 1870, and advertisements for his forthcoming book *The Be-*

ginnings of Life were placed by Macmillan on the front page of *Nature* for many weeks, beginning with the first issue of that year.[21] The subject of the origin of microorganisms was debated at an X Club meeting on 3 February because of Tyndall's recent "Dust and Disease" lecture and an editorial in *Nature* that same day questioning the validity of the germ theory and suggesting that Pasteur's claims about air-germs were just begging the question.[22] Thus, Huxley's interest in Bastian's work must have begun or intensified at this point. "He had been consulted by Dr. Bastian" and was present, along with Busk, Frankland and Sharpey, to witness numerous experiments in March and April 1870, both at the sealing and the opening of tubes when any possible contamination due to experimental error might occur.[23] Furthermore, the support of the publisher Alexander Macmillan for Bastian's work was an important factor in its serious reception at this time. Dr. Gilbert Child wrote to *Nature* to express support for Bastian on 21 April, and Macmillan responded to a complaint on this from Lionel Beale, saying "I think you may rely that Lockyer will allow no partisanship in such questions to influence the fair discussion of them."[24]

Nonetheless, after Bastian's clash with Tyndall in the columns of the *Times* (a forum that would have been inherently distasteful to the X Club for such a disagreement, especially to Hooker, who had advised even Darwin to avoid carrying out such disputes in public), Huxley began to have a more guarded and skeptical view of Bastian's experiments. In particular, he was disturbed when some spiral fibers and a structure resembling a leaf of the moss *Sphagnum* showed up in some of Bastian's sterile tubes, and because the tubes had been hermetically sealed, Bastian was willing to believe that those must have originated from nonliving matter in the solutions. On 1 May 1870, he met with Bastian and cautioned the younger man that he felt such objects must be contaminants. Bastian responded in a letter the next day that he had looked at some *Sphagnum* leaves for comparison and agreed that the one in the solution must indeed be "an accidental fragment of a *Sphagnum* leaf." However, he maintained that the spiral fibers seemed to be genuinely spontaneously generated. He continued: "As I have a good many other solutions now preparing—containing saline solutions only (and which are mostly supposed to contain no carbon) I have quite made up my mind not to say anything at present about the possibility of obtaining organisms without carbon. I hope Dr. Frankland will help me to work this out more carefully. Meanwhile, I shall endeavour to get more information regarding the spiral fibers."[25]

Huxley wrote back, endorsing Bastian's course and, hoping Bastian would not be angry at advice, urging him not to publish until "all results tested."[26] Bastian replied on 12 May after having done more experimental work on the fibers, and seemed to have taken the advice in a collegial spirit: "So far from being at all angry, I am very much obliged to you for the advice contained in your letter, which I know was dictated by the best of motives. I cannot help thinking however that your advice is somewhat severe. I can understand that there is reason for the most extreme caution in bringing before the world supposed new organisms—which may be not organisms at all and not living."[27] However, he felt that some of his experimental results were much more certain and directly contradicted crucial claims of Pasteur, and that those results should be published as soon as possible. Bastian was somewhat puzzled that Huxley's level of caution was so extreme:

> if I get organisms (as in the experiments with Dr. Frankland) in fluids which have been exposed for 4 hrs. to 150°C—this also tends to upset all existing notions. About these experiments there cannot be the least mistake—and they must prove one of two things: either living things can live through such conditions without destruction, or else living things can be produced *de novo*. I cannot see why this should not be made known. All the doubtful part of the enquiry I shall go over again from the beginning. I have already very many solutions prepared which have been exposed to 146°C for four hours.
>
> With regard to the spiral fiber, I am even more strongly inclined to believe it a new growth of some kind. As I mentioned to you yesterday, nearly all the accidental fragments met with in solutions polarize distinctly. I went over a great many specimens in Ladd's shop the other day of vegetable tissues, and the spiral ducts all polarize, whilst not a trace of this reaction is shown by any of my specimens. I have specimens also which seem to show how they became differentiated—and I have now a specimen of the same kind under observation in one of my solutions. Until I am quite certain however, I shall only allude to this as a matter still in doubt.[28]

In the context I have sketched, however, including the recent treason of Mivart, we can easily understand why Bastian's behavior would seem intolerably cocky to Huxley. Bastian was refusing advice from his elder and showing signs that he might eventually resist "eating his leek," and this over the issue of spontaneous generation, which was for Huxley a particularly explosive one for the status of evolution as respectable science.

Huxley Turns against Bastian

Bastian's eagerness to believe in spontaneous generation, and his need to be corrected by Huxley on even something as complex and familiar as a *Sphagnum* leaf, had greatly shaken Huxley's confidence in Bastian's abilities in the laboratory. Thus, even though his tone to Bastian was such that the younger man was still convinced of his support, in a letter to Anton Dohrn at this same time Huxley's tone suggested that he was now much more skeptical of Bastian's claims:

> Then there is our Bastian who is making living things by the following combination:
>
> *Rx* Ammoniae Carbonatis, Sodae Phosphatis, Aquae destillatae quantum sufficit, Caloris 150 Centigrade, Vacui perfectissimi, Patientiae.
>
> Transsubstantiation will be nothing to this if it turns out to be true.[29]

Huxley's attitude toward Bastian may have been still at least a bit ambivalent in early May of 1870 when he seems to have looked in on Bastian's experiments for the last time. However, Bastian went ahead against Huxley's advice and published a large part of his experimental case for spontaneous generation in *Nature* in three installments beginning 30 June 1870. Thus, by midsummer, as Huxley planned and wrote his presidential address for the BAAS in Liverpool and carried out more of his own experiments on the origins of bacteria, molds, and yeasts, he had clearly made up his mind that Bastian's scientific demeanor was improper and unmanageable. In a series of letters exchanged among Hooker, Darwin, and himself at this time, Huxley's tone leaves no more room for any benefit of the doubt toward Bastian.

Hooker opened the exchange by commenting to Darwin that "Bastian's paper in Nature is full of curious matter, but eminently unsatisfactory in treatment. I think it poorly written."[30] Darwin agreed that the paper was not convincing, but continued, "Spontaneous generation seems almost as great a puzzle as preordination . . . Against all evidence, I cannot avoid suspecting that organic particles (my gemmules from the separate cells of the lower creatures!) will keep alive [during boiling] and afterwards multiply under proper conditions. What an interesting problem it is."[31]

Huxley, however, replied to Hooker describing experiments of his own and saying of Bastian's that

> The wonderful and significant fact about Bastian's *Sphagnum* leaves is not that they were in his tubes, but that he had not sufficient histological knowledge to be led to suspect their real nature. He brought a specimen, shut up, to me in order to put an end to my doubts about the generation of living things in his tubes—and I had much ado to convince him of the real nature of the specimens. I believe that I have now made out what his spiral organisms are. The tartrate of ammonia crystals which I am using are full of mycelium of *Aspergillus* and it very readily runs into coiled and spiral form. Furthermore, it does not depolarize light. I put not the slightest faith in Bastian's work. He is a clumsy experimenter and an uncritical reasoner.[32]

And in a letter to Tyndall, Huxley explicitly complained of his concern at having himself been misinterpreted by the public up until now:

> the paper on the Physical Basis of Life was intended by me to contain a simple statement of one of the great tendencies of modern biological thought, accompanied by a protest from the philosophical side against what is commonly called materialism. The result of my well-meant efforts I find to be, that I am generally charged with having invented "protoplasm" in the interests of materialism. My unlucky Lay Sermon has been attacked by microscopists, ignorant alike of biology and philosophy; by philosophers, not very learned in biology or microscopy; by the clergy of most denominations; and by some few writers who have taken the trouble to understand the subject.[33]

Brownian Movement and Other Rhetorical Devices

I have shown in chapter 2 how, between 1828 and 1870, spontaneous generation claims became entwined with theories about particles inspired directly by Robert Brown's active molecules. When Brown died in 1858, his obituary in the *Athenaeum* declared that the cause of Brownian movement was still unknown.[34] In 1868 the well-known Manchester instrument maker and microscopist J. B. Dancer, though attempting to dismiss any and all claims that the particles are self-animated, stated: "The cause of this phenomenon is not yet satisfactorily accounted for. Some have imagined that it is the physical repulsion of the particles when uninfluenced by gravitation. The author has tried many experiments with electricity and magnetism without success. He thinks that the movement may possibly be connected with the absorption and radiation of heat."[35]

In 1863, however, Christian Wiener, professor of descriptive geometry and geodesy at Karlsruhe, claimed to have experimentally investigated and explained the phenomenon. In this paper, Wiener claimed that Brown originally believed that the active molecules might bridge the gap between the living and the nonliving, but claimed himself to have proven that their motion was of a purely physical character:

> Brown . . . believed that this (motion) was a precursor of the continuous life-motions, even in those particles that came from nonliving matter. But this explanation has been given up, and one now assumes that the motion is due to currents that are caused by the never-complete equality of the temperature of adjacent fluid volumes, as well as by constant evaporation. I have now undertaken observations of these motions under the microscope and have concluded that they are caused by the constant motions which fluids undergo on account of their corporeality.[36]

Dancer's theory was not identical to Wiener's, but he likewise made light of any attempts, including those still current, to link the phenomenon with spontaneous generation: "Many instances have come under the author's notice, in which these objects have been regarded by microscopists as animalculae. They have given rise to many very ingenious speculations, some of which are connected with spontaneous generation . . . Some writers who commented on [Brown's] experiments, but who had not carefully followed his communications, asserted that Dr. Brown imagined these particles to be animated—and this statement was generally believed."[37]

By 1870, Huxley had also adopted the belief that the movements were purely physical in character, and this in fact was immediately brought to bear upon his campaign to distance himself from Bastian. Similarly, Brush has noted that "by the 1870s, at least, it was becoming common for authors of books on the microscope to include warnings about

Brownian movement, in case observers should mistake it for the motion of living beings and attempt to build fantastic theories on it."[38] In his "Biogenesis and Abiogenesis" address at the Liverpool BAAS meeting, Huxley began with a very similar tone about Buffon's doctrine of "organic molecules," which he called ingenious for the eighteenth century, but a source of only fantastic whims among those who still held to it in the late nineteenth.

His full-fledged attack on Bastian began a few days later with the delivery at the Liverpool meeting of a paper about Huxley's experiments, "Penicillium, Torula, and Bacteria."[39] It was published soon afterwards. In that paper, Huxley stated:

> When you examine . . . Bacteria with the very highest powers . . . they have *two distinct kinds* of movements . . . These two kinds of movement are not to be confounded. They must be explained as due to very different causes; and it seems to me that it is a confusion of these two which is at the bottom of the mistakes made in the assertions as to the survival of Bacteria, &c., after the application of very high temperatures. I have made experiments with this matter in view. I boiled a solution containing living Bacteria for two hours. On searching for them after this, I found them unchanged in most respects . . . Their life is undoubtedly destroyed . . . everyone admits that; but there they remain with but a slight change of appearance. Do what you will, however, they retain their *trembling* movement; and this is a very misleading phenomenon. Dr. Bastian was good enough to unseal a flask in my presence, which had been closed at a temperature of 150° Centigrade; and I saw there and then Bacteria, exhibiting these active, trembling movements, which, had they come from any other solution, I should have *then* considered as a proof of their being alive . . . The first kind of movement (the trembling) is no doubt the Brownian movement, first shown by Robert Brown to be exhibited by minute particles of a variety of substances, when placed in liquid . . . This discrimination is of the utmost importance. I cannot be certain about other persons, but I am of opinion that observers who have supposed that they have found Bacteria surviving after boiling have made the mistake which I should have done at one time, and, in fact, have confused the Brownian movements with *true living* movements.[40]

Bastian responded to this charge on the spot and repeatedly afterwards, yet found that it was still being used against him as if he had no good defense. In the physicalist view that he still believed he shared with Huxley, it should be noted that he and Bennett (unlike Owen) seemed to fully accept that Brownian movement belonged to the physicists. Indeed, after describing how to distinguish the two kinds of movement, Bastian pointed out that he had even made this distinction *before* Huxley tried to turn it against him, implying plagiarism as well as treachery on the latter's part:

> This statement concerning the two kinds of movements of Bacteria and the power of boiling water to arrest only one of them, is almost word for word what appeared in [Bastian's article in] *Nature* for June, 1870. I thought at the time that the statement was new in certain respects—at least I cannot refer to any similar statement in the writings of others previous to that time. I was somewhat surprised, therefore, on reading the quotation . . ., to find that Prof. Huxley, on Sept. 13, 1870, mentioned such distinctions as if they were quite novel, and with the tacit suggestion that I was unaware of them.[41]

Bastian quoted from Huxley's article and concluded that "what follows is certainly a suggestion that I had been misled by these phenomena, apparently because I was unaware of the distinction then pointed out by Prof. Huxley."[42] The fact that Huxley was able to pull off this rhetorical coup, despite the truth of Bastian's claim about having priority in publishing, is testimony probably not only to Huxley's oratorical skills, but also to his wide reputation in the scientific community, bolstered by much X Club maneuvering for influence in science and in the larger culture. (Huxley, Tyndall, and Hooker had *all* been president of the BAAS by 1874, for example.)

Huxley's Liverpool address was an outstanding success, if measured by the widespread adoption of the terms biogenesis and abiogenesis and their lasting presence in biological literature. Given this, it is more than a little ironic that Huxley had hijacked the term "biogenesis" from Bastian, who was using it up until that time to mean exactly the opposite, i.e. spontaneous generation![43] This must surely rank high in the litany of rhetorical coups pulled off by Huxley in a battle in which defining the terms of the debate was the key to victory. Indeed, we must credit him in this case with even more than his usual degree of rhetorical skill: the fact that biogenesis is often listed in history and biology books today as a term coined by Huxley himself reveals how successfully he denied his opponent the rhetorical high ground, appropriating and permanently reversing the sense of the terms of Bastian's argument.

Bastian attempted to play catch-up by devising his own new terms. He coined the word "archebiosis" to refer to the origin of living matter from inorganic starting materials. In the new scientific naturalist view that Bastian shared with Huxley and Michael Foster, for example, arguing for "spontaneous" generation at any point in the earth's history had become undesirable, as the term "spontaneous" was seen by many to "carry with it the idea of irregularity."[44] Bastian was attempting, as was Huxley with his new coinage, to cut himself off from the older term and its connotations. "Archebiosis" implied *lawful* processes of development, as lawful as, and analogous to, those by which inorganic crystals formed from a saturated solution. Bastian's distinction was favorably recognized by some,[45] though not perceived to have the same elegance as Huxley's "abiogenesis."[46] One reviewer somewhat sympathetic to Bastian's evidence and arguments even went so far as to say, "The nomenclature adopted by Dr. Bastian is very peculiar. The hideously ugly word 'archebiosis' is coined to express an idea, which, when it is examined, is closely allied to that of heterogenesis."[47]

Huxley and Tyndall in their arguments against Bastian continued to describe his views as favoring "spontaneous generation." In the long run, Bastian's failure to separate himself from the implication of randomness was a millstone around his neck. The success of Huxley's "abiogenesis," on the other hand, because it carried the desirable notion of lawfulness, over time helped produce the public perception that Huxley's views were more modern than Bastian's.

Huxley also set a strong precedent in this talk for shifting the discussion of spontaneous generation to the distant past, as required by Darwinian theory. He gained accolades from some for his open-mindedness in declaring that, though he saw "no reason for believing that the feat [of spontaneous generation] has been performed yet," nonetheless "I must carefully guard myself against the supposition that I intend to suggest that no such thing as Abiogenesis ever has taken place in the past . . . That is the expectation to which analogical reasoning leads me; but I beg you . . . to recollect that I have no right to call my opinion anything but an act of philosophical faith."[48]

Huxley was walking his tightrope. This hedging produced outrage, however, among anti-Darwinians and others more totally opposed to spontaneous generation even in theory. One response from this quarter was in William Thomson's BAAS presidential address the following year. Thomson suggested that invoking such atheistic events was unnecessary, because the first life on earth could have been carried to Earth by meteorites from other planets. To Huxley, this was going to laughable lengths to avoid an unpleasant fact. He wrote to Hooker: "What do you think of Thomson's 'creation by cockshy'? God Almighty sitting like an idle boy at the seaside and shying aerolites (with germs) mostly missing, but sometimes hitting, a planet!"[49] Meanwhile, Hooker had written to Darwin expressing puzzlement that Thomson was not inclined to accept Huxley's "in-between" position: "I do not think Huxley will thank him for his reference to him as a positive unbeliever in spontaneous generation—these mathematicians do not seem to me to distinguish between unbelief and a-belief."[50] For his part, Hooker was just as skeptical of meteors as Huxley was: "The notion of introducing life on Meteors is astounding and very unphilosophical . . . Does he suppose that God's breathing on meteors or their fragments is more philosophical than firstly on the face of the Earth? I thought that meteors arrived on the Earth in a state of incandescence—the condition under which T. supposes that the world itself could not have sustained life. For my part, I would as soon believe in the Phoenix as in the meteoric import of life."[51]

To a remarkable extent these discussions anticipate those of the 1990s about possible delivery of organic compounds to earth by comets, and at what point the earth had cooled sufficiently for those compounds to form living things.[52] Despite such immediate reactions as Thomson's, it is increasingly clear from this time forward that almost all scientists sympathetic to evolution began to acknowledge an abiogenetic origin of life, but to allow of this possibility only in the distant past. It is from the months immediately following Huxley's BAAS address that a famous remark of Darwin's finally gave a more explicit and concrete reason for this argument:

> It is often said that all the conditions for the first production of living organisms are now present, which could ever have been present, But if (and oh what a big if) one could conceive in some warm little pond with the right amounts of ammonia and phosphoric salts,—light, heat, electricity, etc. present, thus a protein compound was chemically

> formed, ready to undergo itself such complex changes, at the present day such matter would be instantly devoured or absorbed, which would not have been the case before living creatures had formed.[53]

Bastian's relative youth must have handicapped him in such a dispute among scientific titans. His strategic naiveté was just as important a strike against him. He had cut himself off from a significant traditional support base in the spontaneous generation community (the "molecularists" as Stirling called them) by allowing the physicists to claim Brownian molecules and label as "fantastic" views that the movement was vital. By simultaneously trying to be a good Darwinian evolutionist and Huxleyan physicalist *and* to vocally support the possibility of present-day spontaneous generation, butting heads with Tyndall, Bastian was also alienating the X Club, perhaps not sufficiently grasping its intense desire to be "respectable." Thus, despite his own widely acknowledged oratorical skills, he was soon cut off from those under the influence of the X Club, and had burned the Brownian movement bridge behind him.

Bastian did attempt to respond to Huxley's presidential pronouncements in letters to *Nature* written over a period of a month after the Liverpool BAAS meeting. He raised many of the issues that had previously been raised in private between Huxley and him, and accused Huxley of pretending to have always had clear views that "abiogenesis" was not possible, when not six months previously he had observed Bastian's work with the belief that it might well be proving the opposite.[54] His tone was sharp in response to Huxley's public accusations that his technique was sloppy (a much more high-powered attack than Huxley ever adopted in private when attempting to correct young scientists). Huxley replied with an equally sharp tone, now saying flatly that "what Bastian got out of his tubes was exactly what he put into them," i.e., contaminants.[55] And privately Huxley wrote to Norman Lockyer, editor of *Nature*, to inquire: "I have been obliged much against my will to take notice of Bastian's 'Reply'—What was his reason for going out of his way to be so offensive? He knew exactly what I thought about his work and therefore must have known that in my judgment the kindest thing I could do was to be silent about him."[56]

Bastian clearly did not think that Huxley's public attack at the BAAS constituted a strategy of "being silent about him." And his supporters in the medical community were not daunted by Huxley's attempts to declare him scientific persona non grata. Edward Youmans and Alexander Macmillan also strongly supported Bastian's right to continue the debate with Huxley, and they published Bastian's first book as well as his 1872 two-volume magnum opus, *The Beginnings of Life,* in 1872. Huxley and the X Club, on the other hand, fretted that *Nature* was slipping away from them as had so many previous publishing organs. Now less than a year after the journal's origin, Lockyer and Macmillan's desire to keep it impartial, and thus open to dissenting arguments like Bastian's, threatened to make it a center of divisiveness among the evolutionists rather than the united voice of the X Club that Huxley wanted.[57]

The Younger Darwinians

Once this level of animosity developed, it is not surprising to see among the younger Darwinians such as Anton Dohrn and William Thistleton-Dyer that most soon got the message that Bastian-style evolutionism was incompatible with orthodox Darwinism. In an article in the *Quarterly Journal of Microscopical Science* on 1 October 1870, Dyer wrote, "The interval which the evolutionist is modestly content to conceive deductively bridged, is as nothing to the leaping powers of the so-called heterogenist who boldly widens the gap and passes easily from ammonium tartrate to a Penicillium . . . A believer in spontaneous generation is not really an evolutionist, but is only a vitalist minus the supernatural; the special creation which the one assumes is replaced by the fortuitous concourse of atoms of the other."[58] The need to convince the reader that spontaneous generation was not connected to evolution was a direct indication of the success that Bastian had had with the opposite argument. The sharply negative rhetoric against Bastian is also clear in an 1872 letter from Dohrn to Darwin. Dohrn had already written Darwin about Wallace's "sad falling away."[59] Now a new letter was prompted by Wallace's favorable two-part review of Bastian in *Nature:* "Poor Wallace completely drifts away, and now most unfortunately associates himself with such a man as Bastian! His two articles in *Nature* are the worst thing he ever did in his life,—and it becomes really difficult for his friends to speak with respect of him."[60]

Harvard philosophy professor John Fiske, whose admiration for Tyndall, Huxley, and especially Spencer verged on worship,[61] had at first agreed with Wallace and Youmans that spontaneous generation was still experimentally an open question. In late 1873, while visiting the scientific lights of London, he had written of Bastian's and Huxley's work as illustrations of the nondogmatic nature of evolutionary theory, in a tone that lends Bastian as much authority as Huxley:

> It is perfectly in keeping, for example, for two upholders of the Doctrine of Evolution, as well as for two scientific specialists committed to no general doctrine, to hold opposite views concerning the hypothesis of spontaneous generation. Since this is a strictly scientific hypothesis, . . . invoking no unknowable agencies; and since there is no reason . . . why it should not sooner or later be established or overthrown by some crucial experiment; there is nothing anomalous in the fact of two such thoroughly scientific evolutionists as Prof. Huxley and Dr. Bastian holding opposite opinions as to its merits.[62]

By late 1875 at the latest, however, Fiske too had gotten the message that evolution and spontaneous generation were not to be presented as compatible. His new book, completed in February 1876, contained a noticeably cooler treatment of Bastian:

> When Dr. Bastian tells us that he has found living organisms to be generated in sealed flasks . . ., we demand the evidence for his assertion. The testimony of facts in this case is hard to elicit, and only skillful reasoners can properly estimate its worth. But still it is all accessible . . . and if we find that Dr. Bastian has produced no evidence save such as may equally well receive a different interpretation from that which he has given it, we rightly feel that a strong presumption has been raised against his hypothesis.[63]

As noted earlier, Henry Lawson of St. Mary's Hospital Medical School was an early and vocal supporter of evolution, of heterogenesis in general, and of Bastian in particular. As editor of the weekly *Scientific Opinion* from its beginning in November 1868 until June 1870, of the new *Monthly Microscopical Journal* from January 1869 until his death in 1877, and of *Popular Science Review,* which he took over in January 1869, Lawson wrote frequent editorials in support of Darwinism, pangenesis, and heterogenesis, arguing that all were linked.[64] Many of Lawson's jabs at the orthodox Darwinians, for their resistance to making the logical jump from evolution to spontaneous generation, have been cited already in the previous two chapters. Although these journals targeted a popular, mostly middle-class audience, this was precisely the audience that, thanks to Huxley and others, was becoming the group that practiced science professionally.[65] Lawson and his journals, then, were a big thorn in Huxley's side.

However, Lawson, like the other young Darwinian supporters, also reversed himself within a year or two. In an October 1871 editorial, he was already beginning to question Bastian's claims, with no mention at all of his previous enthusiastic support.[66] And over the next few years, despite the temporary boost that Bastian's work received after Burdon Sanderson's corroboration in *Nature* in January 1873, Lawson's tone became more and more directly opposed to Bastian in all the journals he edited. Indeed, his journals became an important venue for the publication of new evidence from Bastian's opponents and steady editorial criticism of Bastian. By April of 1875, Lawson was reviewing Bastian's latest book, *Evolution and the Origin of Life,* and he now voiced many of the same criticisms as Huxley, Dallinger, and others, especially their belief that bacteria might produce "germs" that could survive temperatures much greater than the adult bacteria ever could.[67] "Dr. Bastian has received many hard blows in this controversy," Lawson noted. Even so, he felt compelled to add, "what is more to his credit, he does not appear anxious to return the blows. Indeed his book is written in the best possible temper, and that is saying a good deal considering the nature of the struggle."[68]

Bastian, Huxley, and the Royal Society

I have discussed Bastian's initial successes at the Royal Society of London. In 1872 Huxley had become the secretary of the Royal Society responsible for biological sciences, replacing William Sharpey.[69] After Bastian's paper on 30 January 1873, Bastian wrote to Huxley in that capacity to ask for the insertion of a footnote to his paper, prior to its printing in the *Proceedings of the Royal Society.* Huxley cordially agreed.[70] However, by the time of Bastian's paper of 1 May, Huxley had begun to exercise the secretary's prerogative to edit papers for the *Proceedings* in a way that Bastian found less to his taste.[71] At the reading of the paper to the Society, Huxley as secretary requested the removal of a note from Bastian's manuscript prior to its publication. The following morning Bastian replied that he had no objection, but went on to request

> At the same time, for the mere sake of the principle involved I should be very glad if you will kindly inform me to what extent it is permissible for one of the officers of the Society to alter any communication after it has been read. Frequently considerable portions of a paper are not actually read at the meeting of the Society, but if omissions of this kind are made for the sake of convenience, is it or is it not the case that the communication as a whole is taken as read? I must say it seems to me rather important that there should be a clear understanding on this subject, although as I said before, I have no objection in the present case to the omission of the "note" to which you called my attention.[72]

This suggests that Bastian already considered the possibility that Huxley would use his position unfairly and was trying diplomatically to secure against that possibility. In a series of letters exchanged over the next two months, Bastian continued to press this point: that as he understood scientific decorum, especially in such an official forum as the Royal Society of London, what had actually been read aloud to the public meeting of the Society must remain inviolate in print unless the author consented to changes.

Huxley, though he refused to say so as a general rule, did not see this principle as sacred, and asserted that greater latitude given to Society officers (such as himself) would best benefit the Society: "Without giving any opinion as to the general question of principles you raise, I may say that I hold myself responsible to the Council for any action which I may take on editing of the Proceedings, and that in the present case I consider myself to be simply carrying out their views as to the manner in which the Proceedings shall be edited."[73]

This of course confirmed Bastian's suspicions, since according to his view of fair procedure this must appear as violating a cardinal principle of respect for another's work, implying possible intent to alter his published words by one who since October 1870 had publicly declared himself Bastian's bitter opponent on the subject of the article. Not surprisingly, then, Bastian became still more guarded and formal in his tone, contributing to the deterioration of his relations with Huxley. For Bastian, Huxley's motive was beside the point: no one but the author should have the right to judge whether a given editorial change might "affect any statement or fact, [or] weaken any argument" he had used. In his next letter he said he had received the revision of his article, but not the proof from which it was made up (Bastian's own original manuscript, presumably), which he wished to compare with the revision to see exactly how great the changes had been. He was told that the proof had been sent to be preserved in the Society Archives. Again stressing his point of principle, he asked Huxley for it, "since without it I cannot readily ascertain what omissions you wish or what modifications you think desirable. As the paper has already been read at the Society, I suppose I am to have some voice as to any proposed alterations."[74] A note in Huxley's hand on the margin of the letter indicates that Huxley directed the proof to be sent to Bastian and had written to tell him so. Both men, though sticking to their principles, seemed to want to avoid conflict.

Once having seen the extent of Huxley's revisions, Bastian offered to assent to some that, he agreed, might be seen as language not sufficiently genteel. He felt, however, that some of the alterations were for more heavy-handed reasons. As he put it: "I am anxious to meet the wishes of the Council—but whilst quite willing to make modifications in the direction indicated, I can scarcely suppose that the Council of the Royal Society would wish to suppress any show of independent opinion . . . The Society as a body is known to disclaim responsibility for any opinions or views expressed by writers of papers."[75] He marked on the revision some changes with which he would be satisfied, "some of the latter having been slightly modified in order to meet what I now understand to be the wishes of the Council" concerning appropriately dignified wording.[76] Bastian emphasized that these were really all the alterations he felt he could fairly be called upon to make, and hoped Huxley would agree that he was showing every sincere desire to find a mutually acceptable solution.

Huxley's reply has not been preserved, but he seems to have balked, for Bastian's next letter is still trying to persuade him that the fact that passages were actually read aloud to the public meeting of the Society ought to guarantee those passages safety from editorial tampering without the author's permission. His earlier expression of confidence that the Council of the Royal Society would definitely agree with him, wishing to avoid any appearance that they might be suppressing dissenting opinion within their ranks, remained intact. He urged more bluntly than ever that if "whole paragraphs or qualifying phrases may be blotted out by the editor . . . even in the absence of any intrinsic unsuitability in the passages themselves . . . This, . . . I venture to think, would involve a great departure from existing rules."[77] But Bastian did not feel that all hope of respectful interchange was gone. He closed by repeating that, "having myself freely consented to several omissions that have been proposed, I trust I may be allowed to have an opinion as to other modifications. Encouraged however by certain expressions in your last letter, I hope when you find that the passages marked . . . were actually read to the Society, you will not still think it necessary to cancel them, in opposition to my own wishes."[78]

Huxley continued to wrangle over the question of whether the passages actually had been read, the two men comparing differing recollections on this subject until it became almost comic. Finally Bastian gave up hope of avoiding a confrontation and begged

> to state that only one consistent course seems open to me . . . I can scarcely think that the Council will insist upon the removal of the words and passages in question . . . It seems to me moreover highly desirable that Fellows should know whether they are at liberty to state in any communications that they make . . . that a particular hypothesis under discussion is "tenable" or "untenable," etc. Should the Council desire to impose restrictions against expressions of opinion of this kind, I, of course, must bow to their decision and consent to the publication of my paper in the curtailed form . . . I am therefore quite content that its publication should stand over until the Council shall have come to their decision upon this subject.[79]

Bastian seems to have had faith in his interpretation of the rules of fair play, and in the Council's desire to avoid negative publicity. But he was hedging by trying to guarantee that if he was to be censored, at least it would have to be done by an official act, so that other members of the Society would know that suppression against his wishes had occurred and could happen to any of them as well.

Huxley, on the other hand, seems to have had no doubt that the Council would fully support his position in the matter. The conflict must be viewed in the context of the X Club's consolidation of power in the Royal Society during these years, so thoroughly explicated by Ruth Barton. Because the X Club had at least three members on the council at all times during the 1870s (five during 1873)[80] and already exerted wide influence on Society politics by the time of the dispute, Huxley knew he could rely upon a friendly jury. Perhaps more important, secretaries of the Royal Society during the nineteenth century were in general granted enormous latitude by the Council and thus had considerable personal power. Huxley was no exception; he influenced who received Royal Society research grants, for example, to the point where one historian notes: "If Huxley's intentions were honourable and his reply fair, it remained clear that the Government Grant had been given by the few to the few."[81]

Thus, Huxley graciously agreed to Bastian's suggestion, perhaps himself despairing that the younger man did not understand the realities of the situation in his refusal to accept Huxley's advice and generosity. Perhaps Huxley had been trying to handle things without such a confrontation, which he knew would likely lead to an embarrassing defeat for Bastian.

In the event, the Council did indeed fully support Huxley. Further, they did not even admit to seeing Bastian's concerns as reasonable ones. They advised Huxley to communicate to Bastian "That the Council see no reason to interfere with the action which has been taken by the officer in the matter of Dr. Bastian's paper." Huxley did so and offered an olive branch to the loser: "I hope that you will believe that nothing but a sense of my duty to the Royal Society has led me to exercise my editorial functions in a way which, I fear, has been disagreeable to you and that I have done my best to avoid the suggestion of any omission which would really tend to weaken the strength of your arguments against those to whose views you are opposed."[82] Nonetheless, it was very clear to Bastian that the Royal Society was no longer a fully supportive nor in his view fully objective venue for his scientific papers. He began to explore other possible forums, such as the Pathological Society of London.

E. Ray Lankester and Bastian

Meanwhile, Henry Acland, Regius Professor of Medicine at Oxford, communicated a paper to the Royal Society on 19 June 1873 by two of his young colleagues, C. C. Pode and E. Ray Lankester, criticizing Bastian's experiments on various technical grounds.[83] It should be recalled that Lankester was a devoted disciple of Huxley, having been one of the demonstrators who helped out with the elementary biology course Huxley had just begun to teach at South Kensington in the previous two summers. Huxley had also arranged for Lankester to work at the Naples marine station that Anton Dohrn was just beginning, and Lankester and

Dohrn became fast friends between 1871 and 1873. Dohrn's scornful views on Bastian have already been alluded to. Lankester was a very eager proponent of abiogenesis in the distant past producing *Bathybius,*[84] and he went on to supervise the English translation of Haeckel's *Natürliche Schöpfungsgeschichte* from 1874 to 1875. Huxley supported Lankester as an ardent disciple, very talented in microscopic work. The young man had taken over from his father Edwin Lankester the editorship of the *Quarterly Journal of Microscopical Science* in 1869 at the extraordinarily young age of twenty-two. Still, though Dohrn thought Lankester "one of the best of the young generation of English biologists," Huxley replied that he had a vague anxiety that Lankester was unstable: "I should like to see him do well, but there is what we call 'a screw loose' about him. I don't know exactly what screw it is, but there is something unstable about him."[85]

Lankester had begun his public attack on Bastian in the January 1873 issue of the *Quarterly Journal of Microscopical Science.* Perhaps validating Huxley's estimation of his judgment, Lankester's first salvo was unusually intemperate for a scientific review, calling Bastian a "mesmerized victim of delusion." He went on to say, "The origin and mode of growth of such delusions form a very interesting psychological study, and it is only when we have obtained a proper conception of Dr. Bastian as an abnormal psychological phenomenon that we can hope rightly to appreciate the whole of statements made in his book."[86] This was a rather strident tone for a twenty-five-year-old to take with a man ten years his senior and of comparably greater scientific standing.

Still, by summer of 1873, at the same time his paper was being read to the Royal Society, Lankester was stepping up his use of the *QJMS* as a forum for attacks on Bastian and spontaneous generation. By the fall issue, he had run an article by Joseph Lister (that also, incidentally, agreed with his own pleomorphist position advocated in the same issue). Joseph Hooker's lieutenant, William Thistleton-Dyer, wrote to Lister in advance to discuss the latest evidence of Lankester and Pasteur and to ask for a reprint of the forthcoming paper.[87]

When it came to Huxley's code of behavior for young scientists, Lankester excelled, especially in deportment toward his superiors (the X Club) over points of disagreement. His behavior here contrasted most sharply with that which had made Bastian such an outcast, despite the fact that Huxley and others considered him personally often just as difficult as Bastian. One episode from 1875 illustrates Lankester's willingness to be a team player, if only for the team of his revered Huxley. Lankester had concluded from his own anatomical studies on *Amphioxus* that Huxley had made an error concerning the development of certain lateral folds in the embryo. Huxley had used too strong an alcohol solution as a preservative, and this had caused excessive contraction and folding. Lankester usually showed great zest for attacking mistakes of others in print.[88] In a letter to Dohrn, Lankester described how he planned to deal with the situation in this case: "Huxley and Wilhelm Müller are both quite wrong . . . due to their having worked with specimens preserved in strong reagents, especially absol. alcohol which contracts the whole affair. I tell you this especially because I am not going to publish it—as I don't like to be the agent for setting Huxley right. I want to show him the things and get him to publish it himself."[89]

Lankester was also, heart and soul, devoted to Huxley's project of reforming English medical and physiological education, and of rooting both firmly in a course in elementary biology. To say that Lankester was more devoted to Huxley than to Acland is an understatement. Acland bragged about biology and medical education at Oxford at this time,[90] but Lankester in private called him a pompous, out-of-date fool and loudly bemoaned the dismal state of biology, especially laboratory work, in a letter to Huxley:

> A commission comes and examines people such as Acland, and of course hears the smoothest, most cheerful account of everything. But many of his statements are untrue and meant to deceive—as for example when he speaks of his having so much work to do in the University—when it is notorious that he does nothing, and though Professor of Medicine, never gives a lecture and tries to prevent medical students from coming to Oxford . . . So, until there is some movement, in accordance with the recommendation of the Commission . . . I am simply kept to my present position, a mere College advertisement—"We have a Natural Science lecturer on our staff" . . . I should like any to be undeceived who might help me to get away from the position of being a helpless fettered spectator of chaotic want of organization, and complacent optimism.[91]

Lankester was not one to mince words when it came to an opponent. Even in Huxley's most pointed letters one rarely sees this kind of frank scorn for a person in authority. Lankester's bluntness was to make him enemies, including Burdon Sanderson and Sharpey-Schafer, with whom he might have been expected to be allies (as Huxley was). In the spontaneous generation debates, Lankester was not inclined to be as genteel as Huxley in dealings with his opponent. When he obtained the very first negative results from the experiments to challenge Bastian's claims, he referred to Bastian in very disparaging terms. He claimed he had repeated some of Bastian's key experiments, with negative results, and would soon be publishing his own work to finish Bastian off handily.

Unfortunately for Lankester, this was just a week before Burdon Sanderson announced in January 1873 that he had succeeded in replicating Bastian's results on the very same experiments. Having raised the level of personal invective in the debate, he was treated to a lesson in logic by Bastian, who taunted:

> perhaps [Lankester] might with advantage also reflect a little more closely upon . . . the negative results to which he is so fond of alluding . . . Not long ago Mr. Lankester, upon the strength of his own negative results, triumphantly announced that he was about to prove to the world the falsity of my views, and so help to justify the opinion which he at the same time expressed as to my being "the mesmerised victim of delusion," "an abnormal psychological phenomenon," and many other fine things. But unfortunately for Mr. Lankester, just about the same time Dr. Sanderson (whose opinions he so much respects) . . . helped to show, in fact, that my positive results were worth more than the many negative results obtained by other workers.[92]

Tensions within the community of evolutionary scientists were an important part of the story of the debate in Britain in the 1860s and 1870s. The ambiguity of whose experiments were the reliable ones was not resolved until the dust settled over Darwin's indecision about spontaneous generation, when sympathizers realized that supporting Darwinian evolution was only possible by supporting Huxley and Tyndall's version. In other words, scientists came to the conclusion that Huxley and Tyndall spoke for Darwin, and renounced the previous appeal of Bastian's arguments. As we will see in chapter 7, this consensus also coincided with Tyndall's introduction of heat-resistant spores into the debate, so that it no longer became necessary to see Bastian as incompetent or deliberately deceitful in order to conclude that he was mistaken. Huxley himself seems to have decided that Bastian was too naive and self-willed to be a reliable lieutenant in the Darwinian ranks very early on, before the experimental evidence was at all conclusive. I have also shown the importance of Huxley's success in defining the terminology of the debate, and that Bastian's failure to popularize his own terms gradually undermined his support, even among evolutionists sympathetic to his logic. Furthermore, Huxley's outflanking of Bastian at the BAAS in 1870 and at the Royal Society in 1873 demonstrated his effective use of his rank in both organizations to undercut Bastian's position.

5

Colloids, Pleomorphic Theories, and Cell Theories: A State of Flux

Thirty years ago, John Crellin wrote,

> I do not think it would be inaccurate to label the medical and biological world of the 1860s, "the era of the particle": one reads, for example, of biads, bioplasts, gemmules, germs, grafts, globules, oleo-albuminous molecules, morbific ferments, physiological units, and viruses. While some of these particles were essentially concerned with theories on the development of tissues, on heredity, and on the spontaneous generation of microorganisms, they nevertheless provided biological background which was not infrequently called upon in the discussions on infection and spontaneous generation.[1]

This remark, and similar comments that investigators in the 1860s "were obsessed with the novelty of the supposed life-substance . . . called protoplasm," would have us believe that the attitude of mid-Victorian scientists is best viewed with Whiggishly patronizing amusement.[2] What I hope to show here, however, is the extent to which the entire theoretical problem complex associated with the spontaneous generation debates can only be understood in its full, interwoven complexity by taking the "particles" and "life-substance" very seriously.

From 1860 to 1880 theoretical issues related to the spontaneous generation controversy were in a state of active change and development. The controversy shaped those changes in critical ways and was dialectically affected by them. In this chapter I argue that the closure of a number of these theoretical issues removed crucial pillars of support from the advocates of spontaneous generation, much as the shifting of the origin of life question exclusively to the distant past by Huxley and the orthodox Darwinians closed off a significant theoretical space for spontaneous generation views and their linkage to the theory of evolution. During the period 1860–1880 the spontaneous generation doctrine, along with other doctrines such as the correlation of forces and the protoplasmic theory, was seen to be directly relevant to the larger question of the origin and meaning of "vitality." The meaning of vitality in Victorian culture was still fraught with religious content. In this chapter and the next I will show that the theological implications of spontaneous generation were thus never far from the surface, as had always been the case throughout the history of the doctrine. That molecular doctrines had immediate implications for these larger issues is clear from treatises by respected authorities in other scientific fields; in chemistry, for example, on the topic of diffusion.

Thomas Graham and Colloids

A topic closely related to "histological molecules" in the 1860s and 1870s was Thomas Graham's analysis of colloids. Indeed, for Bastian and many spontaneous generation supporters of the 1870s, colloids assumed a role in their arguments very much like that formerly held by the now outdated "histological molecules." Graham had been a professor of theoretical chemistry at University College London since 1828 and had trained an entire generation of scientifically and mathematically minded Victorians, among them Henry Bence Jones, William Stanley Jevons, and Bastian.[3] He was considered "one of the great pioneers of chemistry through his work on molecular motion."[4] Graham had been an early and consistent supporter of Liebig's model of infectious disease, which called for fermentation by chemical catalysis, arguing particularly that the contagious matters were not likely to be gases, but rather complex organized particles.[5] Graham had also become the highly respected Master of the Mint and was nearing the end of his long teaching career when he summed up his decades of research on colloids in a widely cited paper in *Philosophical Transactions of the Royal Society* in 1861.[6] Here Graham argued for a fundamental dichotomy in chemistry, between what he called colloid compounds and crystalloid ones. Colloids were distinguished by their extremely low rate of diffusion, absence of the power to crystallize, gelatinous hydrates, and inertness (as acids or bases and in most chemical reactions). Graham argued that these seemingly inactive properties did not make colloids unimportant. Quite the contrary: "the . . . chemical indifference referred to, appears to be required in substances that can intervene in the organic processes of life. The plastic elements of the animal body are found in this class. As gelatine appears to be its type, it is proposed to designate substances of this class as *colloids*, and to speak of their peculiar form of aggregation as the *colloidal condition of matter.* Opposed to the colloidal is the crystalline condition . . . The distinction is no doubt one of intimate molecular constitution."[7]

Graham was referring to the molecules of the chemist (not the histologist) when he spoke of "molecular constitution." Though his article is a lengthy report on the diffusion of colloids, he continues in a rather more speculative vein to a remark which, though brief, was nevertheless one of the most widely cited ideas in the entire paper: "The colloidal is, in fact, a dynamical state of matter; the crystalloidal being the statical condition. The colloid possesses ENERGIA. It may be looked upon as the probable primary source of the force appearing in the phenomena of vitality."[8] Graham's dualistic distinction was taken up very widely within a very short time in scientific and medical circles.[9]

Benjamin Ward Richardson, noted researcher in experimental pharmacology and advocate of the Liebigian chemical theory of contagious disease, quickly adapted Graham's terminology to that approach when he said, "The physical theory of the origin of communicable diseases is that the diseases are due to poisons which are organic . . . It accepts that the organic poisons may be colloidal and may be transmitted . . . on solid substances, or in water, or by air; but it holds that their action is purely physical on the body."[10]

Henry Maudsley, a young doctor specializing like Bastian in neurology, discussed Graham's paper in a highly influential review article in 1863. In that review Maudsley embedded colloids in the much larger problem complex outlined in this chapter and the next.[11] This is evident in perusing the group of works that Maudsley included in his review: when we look at Virchow's *Cellular Pathology,* Graham's colloids, Matteucci's animal electricity, and other recent works on the electrical nature of the nerve impulse, we might see a hodgepodge of unrelated topics. To see the connections that were apparent to a Victorian reader, we must reconstruct the large problem complex that Maudsley outlined.

Maudsley's article "The Theory of Vitality" stated that vital force is merely the most complex of a series of interchangeable forces such as heat, electricity, "chemical force," and so on. One might compare the ranking of these forces to the differing complexity between organic compounds of tissues and simple inorganic compounds. Because of the complexity of the vital force, which he frequently refers to as its "greater dignity" (the moralizing nature of his analysis of forces is evident throughout, and is typical of mid-Victorian science, especially when discussing "nerve force"), he argued that the vital force must have come along later by an evolutionary process, just as organic compounds can only have evolved after and from simpler inorganic ones.

Maudsley set up a duality of dynamic processes: attraction (gravitation, molecular cohesion, magnetism) versus repulsion (the "centrifugal force," heat, electricity) that he said is analogous to the duality in matter (colloid versus crystalloid). He argued, along with Graham, that because there is interconversion of forces, it should also not be surprising to see some substances that can make the transition between the colloid and crystalloid state. Maudsley then maintained that change of *force* (for example, by nerve discharge or muscle work, producing heat, electricity, and so on) always "coincides with a molecular change in the statical element." For instance, in nerves and muscles acid is produced, and after exhaustive mental activity, phosphate is produced. Both must be restored by nutrition during rest in order for the corresponding high energy states to be reestablished. "The cell thus . . . preserves its individuality; and definiteness of energy, with the maintenance of individuality, are what is connoted by 'vitality' . . . The colloidal is a dynamical state of matter, the crystalloidal a statical state. The colloid exhibits energy, its existence is a continued metastasis."[12]

Maudsley also included a lengthy discussion of how states of disease correspond to a lower vital energy. In inflammation, "the injury has so damaged the parts that the vital force cannot rise to its specific elevation; an inferior kind of vital action alone is possible, which is really disease."[13] Much Victorian morality is embedded in the argument that "quiet, self-animated activity" is *higher* than the "unrestrained dissipation of lower activity," for example, heat in inflammation or emotion-laden displays of force. The latter is contrasted with "the man [meant in a definitely gendered sense] who has developed emotional force into the higher form of will-force, who has coordinated the passions into the calm, self-contained activity of definite productive aim."[14] Thus he classifies epilepsy, and emotional states in general, as "lower" states, meant both quantitatively as well as moralistically.

Maudsley claimed that, because of the difference in level of organization, "one nervous unit, monad, or molecule is the vital equivalent of many units, monads, or molecules of polype substance."[15] And from this he drew an analogy between a complex metazoon and a complex political state. This use of the latest in correlation of forces and "molecular" science as the basis for a scientifically founded theory of behavior, politics, and so on was not peculiar to Maudsley among Victorian thinkers of his generation. His contemporary William Stanley Jevons maintained a lifelong interest in ultimate particles of matter, their Brownian movement, and the laws of their attraction and repulsion. For Jevons also, these served as more than mere analogies for his attempt to construct a theory of economics based on the unit of the individual person and his pleasure and pain ("attractive" and "repulsive") motivators.[16] Jevons discussed Brownian movement in particular with microscopist John Benjamin Dancer and chemist and sanitarian Robert Angus Smith between 1868 and 1870 at the meetings of the Manchester Literary and Philosophical Society, which all three men regularly attended.[17]

The meaning of "colloid," especially in regard to its implications for spontaneous generation and even larger questions, was a subject of dispute. Graham's student Jevons argued that Brownian movement was *not* living in character, but was "closely connected with the phenomena of osmose so fully investigated by . . . Mr. Graham."[18] Bastian emphasized the importance of the colloid concept, both in his very first paper in *Nature* in June 1870 and in his theoretical formulations. Bastian argued that the fact that inorganic and organic substances could both exist in the colloid state, and thus both possess "energia," showed that a continuum of particles across the organic/inorganic boundary already existed; particles that had the necessary property to form living things. He developed this idea further in the theoretical introduction to *The Beginnings of Life,* also implying that such a continuity in nature was called for by the doctrine of the correlation of forces.

Benjamin T. Lowne, a member of the Royal College of Surgeons, was skeptical of the possibility of abiogenesis, yet in an address to the Quekett Microscopy Club after the 1870 Liverpool BAAS meeting, he still felt the need to qualify his skepticism by adding, "Perhaps, as Dr. Bastian suggests, colloid may be intermediate between inorganic and organic living material . . . No doubt, with Mr. Charles Darwin's hypothesis, the origin of living organic from inorganic matter would supply a gap in the evolution of the animal kingdom."[19]

Colloids, then, could do much more work than mere diffusion chemistry. One's view on the nature of colloids had constant large implications for issues from politics, to economics, to the meaning of life and its distinction from the nonliving, to evolution. The minimizing of Graham's original claim, in order to place colloids clearly within the realm of physical sciences, was part and parcel of the same treatment for Brownian movement and the separation of both from the soon-to-be defunct doctrine of histological molecules.[20] Because Bastian had so clearly tied his case for spontaneous generation to the views on colloids and on "molecules" that were about to become obsolete, the defining of colloids as a purely physical phenomenon removed a pillar of support for spontaneous generation.[21]

Conservation of Energy and Correlation of Forces

As shown by Maudsley's 1863 review article, the brief speculative remark in Graham's paper on colloids pointed to a much larger theoretical complex to which his colloid/crystalloid distinction was perceived to be related: nothing less than "the theory of vitality." The theory of vitality included the origin of the forces involved in consciousness and mental activity. This larger problem complex was connected to another idea, the so-called "correlation of forces." This was the name current in the 1860s for the principle that heat, electricity, light, and all other physical forces are interconvertible, a process in which the overall quantity of force or energy is conserved. It was also believed that the energies at work in living organisms were fully interconvertible with those purely physical energies, also involving quantitative conservation. This broad doctrine had been independently developed by Robert Mayer, Hermann Helmholtz, and William R. Grove, among others.[22] An attempt to work out the nature of the linkage among these doctrines could also be seen in the work of Henry Bence Jones in 1867 and 1868.[23] A trainee of Liebig's laboratory, like Graham, Bence Jones supported Liebig's view of disease into the late 1860s, even in the face of Pasteur's experiments.

Richard Owen was also trying to make sense of the relationship between physical forces and the forces in living organisms. His writings make clear that even more broad scientific concepts were involved. It has been said that Owen's stand in favor of heterogenesis failed to win a wider audience because he wrote in an extremely convoluted style. While this claim has some validity (Owen's writing can be exasperatingly convoluted and fuzzy), it has nonetheless contributed to a failure to understand the way in which Owen was also trying to address the larger issues of the day. His use of the term "centres of force" in reference to atoms (see chapter 2) indicated a specific stance Owen was taking on several related and hotly debated questions: Were atoms real material objects as opposed to mere heuristic creations useful in physics and chemistry? Was matter ultimately inseparable from force, or were the two distinct entities? Was an "aether" needed to account for light and other electromagnetic waves?[24] In Britain by the 1860s, the discussion of all these issues had become part and parcel of the discussion of that same larger idea "first represented by the term conversion of force, then by correlation of forces, then by conservation of force, and now by conservation of energy."[25] It was essentially an anti-vitalist agenda, arguing that there was quantitative interchangeability and conservation of vital and physical forces, i.e., that the forces animating living matter obeyed the same laws as purely physical forces and were not separated from those by any fundamental gulf. For most British scientists (including the physiologist William Carpenter and the members of the X Club) this did not necessarily imply the possibility of abiogenesis; indeed, "force" could be regarded as synonymous with the divine. For William R. Grove, the most prominent British founder of the concept of correlation of forces, essential causes still remained unattainable and Creation still the act of God.[26] In the view of a minority, including Bastian, however, the correlation of physical and vital forces implied that there was no great insurmountable gap to cross in order for nonliving materials to organize themselves into living matter.[27]

Michael Faraday's work and writings were considered seminal on many of these questions about atoms, aether, and force, and Owen, Bence Jones, and Tyndall all claimed Faraday as the source for their ideas in the year just after Faraday's death, 1868.[28] Public acclaim for Faraday was at its peak during this period: many claimed him as mentor and sought to gain credibility by basking in the authority of the master. Indeed, Owen's argument for what he calls a Faraday-like interpretation of atoms as "centres of force" could have been written as a specific reply to

Bence Jones's 1868 Croonian Lectures to the Royal College of Physicians, in which Bence Jones had urged that "[a]s soon as it is admitted that force is absolutely inseparable from matter . . . it will become as impossible to think that matter can consist only of centres of force, as to think that matter can be inert or void of all force."[29]

The respected London instrument-maker John Browning believed that the doctrine of correlation of forces implied that the transition from living to nonliving might not be an impassible barrier. In 1869, when Henry Lawson was still an avid supporter of spontaneous generation, he published an article to that effect by Browning in two of his popular science journals.[30] The correlation of forces doctrine was also a very important part of Bastian's overall argument for a context in which spontaneous generation made sense, from his earliest writings on the subject, including the *British Medical Journal* series of 1869. He emphasized its importance again in the lengthy introduction to *The Beginnings of Life,* in which he stated that the modern correlation of forces doctrine implied continuity in nature across the living/nonliving boundary.[31] Bastian apparently convinced many educated people. For instance, Frederick Barnard, president of Columbia College in New York, wrote in 1873:

> To the philosopher, the demonstration of the theory of spontaneous generation . . . cannot but be a matter of the deepest interest . . . Nor is this the only theory which the investigators of our time are urging upon our attention, of which I feel compelled to make the same remark. There are at least two besides which impress me with a similar feeling, and the three together constitute a group which, though to a certain extent independent of each other, are likely in the end to stand or fall together. These are the doctrine of spontaneous generation, the doctrine of organic evolution, and the doctrine of the correlation of mental and physical forces.[32]

Nor was this statement Barnard's only interest in this complex of ideas. Barnard, like many others, saw the real stakes behind a scientific worldview that extended to explaining human consciousness—nothing less than a final choice between materialism and religion. He pointed out the larger and more threatening implications of such a worldview, namely that "[i]f these doctrines are true, the existence of an intelligence separate from organized matter is impossible, and the death of the human body is the death of the human soul. If these doctrines are true, the world becomes an enigma, no less to the theist than it has always been to the atheist."[33] Grove, Carpenter, and Huxley had been at pains for years to find a way to blur such implications of scientific naturalism (note Huxley's famous coinage of "agnostic" for example) as part of their larger project to establish a voice of cultural authority for science and to make it a viable profession. If their position amounted almost to a kind of Carlylean romantic pantheism, Huxley and Tyndall had no qualms in preferring that to naked materialism (though their disavowals of materialism never openly admitted to such a pantheism). As a result, they were accused of being disingenuous by most of their critics, who saw "the Physical Basis of Life" or Tyndall's 1874 Belfast address as direct statements of materialist philosophy.[34] Even many sympathetic to evolution could not see a way out of the dilemma, and Barnard was one of these. If he was rather more explicit about his distress than was consistent with the larger project of Huxley and colleagues discussed earlier in this chapter, perhaps his location in New York, somewhat out of the mainstream of London rhetoric, can excuse him. His exceptional frankness bears quoting at some length:

> We are told, indeed, that the acceptance of these views need not shake our faith in the existence of an Almighty Creator. It is beautifully explained to us how they ought to give us more elevated and more worthy conceptions of the modes by which He works His will in the visible creation. We learn that our complex organisms are none the less the work of His hands because they have been evolved by an infinite series of changes from microscopic gemmules, and that these gemmules themselves have taken on their forms under the influence of the physical forces of light and heat and attraction acting on brute mineral matter. Rather it should seem we are a good deal more so. This kind of teaching is heard in our day even from the theologians [Stebbing, Kingsley] . . . It is indeed a grand conception which regards the Deity as conducting the work of His creation by means of those all-pervading influences which we call the forces of nature; but it leaves us profoundly at a loss to explain the wisdom or the benevolence which brings every day into life such myriads of sentient and intelligent beings, only that they may perish on the morrow of their birth. But this is not all. If these doctrines are true, all talk of creation or methods of creation becomes absurdity; for just as certainly as they are true, God himself is impossible . . . But we are told it is unphilosophical, in the pursuit of truth, to concern ourselves about consequences. We should

accept the truth with gladness, . . . and let consequences take care of themselves. To this canon I am willing to subscribe up to a certain point. But if, in my study of nature, I find the belief forced upon me that my own conscious spirit . . . is but a mere vapor, which appeareth for a little time and then vanisheth away forever, that is a truth . . . for which I shall never thank the science which has taught it me. Much as I love truth in the abstract, I love my hope of immortality still more; and if the final outcome of all the boasted discoveries of modern science is to disclose to men that they are more evanescent than the shadow of the swallow's wing upon the lake, . . . give me then, I pray, no more science. Let me live on, in my simple ignorance, as my fathers lived before me, and when I shall at length be summoned to my final repose, let me still be able to fold the drapery of my couch about me, and lie down to pleasant, even though they be deceitful dreams.[35]

It is not surprising that many, including Alfred Russel Wallace, would eventually choose the same way out of the dilemma that Barnard chose. It is from this time that Wallace began to keep his growing interest in spiritualism in one airtight compartment and his evolutionary ideas in another, drastically reversing his previous summer's enthusiasm for Bastian.[36] Thus, another pillar supporting spontaneous generation began to wobble.

Cell Theory and the Demise of Histological Molecules

As already noted in chapter 2, Huxley's well-known paper "On the Physical Basis of Life" described granular protoplasm in a way that was interpreted by many to confirm their views on "granules" or "molecules" of vital matter. Lionel Beale argued with Huxley's terminology and tried to impose his own term, "bioplasm," on the particles of vital substance, whereas John Hughes Bennett, Henry Lawson, Ernst Haeckel, and numerous others believed the granules of protoplasm to be the same as Darwin's "gemmules" or "living atoms."[37] All of them, however, whether critics or fans of Huxley, believed he was arguing that primitive granular protoplasm was the intermediate step between the living and the nonliving. Thus, notwithstanding Huxley's protests that he intended no such consequence, in 1869 and 1870 the sudden popularity of this recent addition to cell theory gave a major boost in Britain to the fortunes of spontaneous generation. Lawson pointed out the similarity with Bennett's "molecular theory" of cell structure repeatedly and at length.[38]

Like "colloids" and "protoplasm," "molecules" had now also come under dispute. As late as his 1872 *Textbook of Physiology* Bennett still used the same terms as in 1863 to describe his molecular theory. It will be remembered, however, that Beale had criticized Bennett's terminology. Beale said of the terms "granule" and "molecule," for instance, that "much confusion has arisen from the use of terms which have not been well defined." By 1864 Beale attempted to substitute "globule" for "granule," and went on to add (in a clear reference to Bennett): "So, again, the term 'molecule' has been employed in some cases synonymously with 'granule,' but it would obviously be wrong to speak of a small globule as a molecule. It seems to me very desirable to restrict the terms 'granule' and 'molecule' to minute particles of matter which exhibit no *distinct structure* when examined by the highest powers at our disposal, and the term 'globule' to circular or oval bodies of all sizes which have a *clear centre,* with a *well-defined dark outline.* Other examples of the use of insufficiently defined terms might be pointed out."[39] Much as in the rhetorical contest between Bastian and Huxley, Beale wanted to impose his definitions of the terms because they contained his vitalistic assumption that bioplasm particles could come only from previously living tissues.

In an 1872 reprint of his "Dust and Disease" lecture, John Tyndall continued his opposition to spontaneous generation and he also singled out Bennett for criticism, as one of the most widely respected advocates of that doctrine. It is noteworthy, however, that while Tyndall criticized the conclusions Bennett drew from his experiments, the physicist took pains to explain that he was not challenging Bennett's competence as an experimenter or the results he had obtained.[40] Nonetheless, the fortunes of Bennett's theory were soon to decline rapidly.

By this point, Bennett's failing health led him to spend more time in the south of France, and his frequency of publication was greatly reduced. However, late in 1872 his lectures on histology and physiology appeared in textbook form. It was clear from this latest version of Bennett's views that his position had changed little: he still held that the molecule was primary to the cell as a unit of structure and function and he advocated heterogenesis as strongly as ever.[41]

In mid-1873, a reviewer of Bennett's 1872 *Textbook of Physiology* expressed surprise that Bennett could still propound ideas on cell theory that the reviewer considered decades out of date. In just the short time since his widely acclaimed 1868 paper, the small but still respected minority of cytoblastema advocates had shrunk to almost a minority of one. The general stature of Darwin might have been expected to slow this process, since he had spoken with respect of that minority school of opinion in his initial description of his pangenesis hypothesis. But support for Darwin's hypothesis had not materialized, and pangenesis was widely considered to have received a grave blow from recent experiments by Francis Galton. Darwin's gemmules received what many scientists considered another serious, even fatal blow in 1875, from James Clerk Maxwell in his article on the "Atom" in the ninth edition of *Encyclopedia Britannica.*[42] Maxwell argued that since one of the histologists' "molecules," or Darwin's gemmules, could contain at most about one million organic molecules, it was impossible to imagine it as the germ encompassing the full hereditary potential of a higher organism, complete with all its organ systems. Thus, as Darwin's views on this question were considered wrong by most, so Bennett's more extreme "molecular" theory was regarded even more skeptically.

Furthermore, the 1872 version of Bennett's theory was not so different from the theory as he had published it in 1861 and taught it since the early 1850s. In the interim, most biologists had become converts to Virchow's *omnis cellula* doctrine, while Bennett went on believing that he had dealt Virchow's doctrine a crushing blow by 1861 (just as he felt he had quelled Beale's more extreme form of vitalistic critique in 1863).[43] Bennett thus did not attempt to respond further in his 1872 text to more recent cytological and histological evidence. He also took a very similar stance toward Pasteur's experimental work on spontaneous generation, as in his initial 1868 critique of it. Thus, to his reviewer Cleland in 1873, it appeared that Bennett was simply ignorant of the new evidence and the changing tide. Furthermore, Huxley's public admission in mid-1875 that *Bathybius* was a chemical artifact rather than a primitive form of living protoplasm finally undermined the linkage between support for the protoplasmic theory and support for spontaneous generation.[44] In addition, Huxley tried in his popular lectures on science to clarify the confusion between the older histological molecules and the physical ones with which biologists should now be concerned. Twisting around Maxwell's argument, Huxley also sought to clarify the distinction by emphasizing the difference in size. When writing of the granules described in a very small monad, for example, he said, "The authors whom I quote say that they 'cannot express' the excessive minuteness of the granules in question, and they estimate their diameter at less than 1/200,000th of an inch . . . Nevertheless, particles of this size are massive when compared to physical molecules; whence there is no reason to doubt that each, small as it is, may have a molecular structure sufficiently complex to give rise to the phenomena of life."[45]

By the time of Bennett's death in September 1875, even a very sympathetic obituary in the *British Medical Journal* could say only: "He contributed to physiology his well-known molecular theory of organisation which, if not adopted generally, has at all events led to a reconsideration of the question by the upholders of the cell-theory, and to considerable modifications in their views."[46]

The spontaneous generation controversy was still raging in Britain at the time of Bennett's death, and it may well have been his unbroken opposition to both Virchow and Pasteur that first led their work in such widely differing fields to be considered intimately connected in biology textbooks. Virchow himself had begun a renewed campaign shortly after Bennett's death to discredit the remnants of belief in Schleiden and Schwann's cytoblastema theory. He now argued that one of the greatest problems with that view was precisely that it left the door open for, indeed even implied, spontaneous generation of cells.[47] This was an active historical reconstruction by Virchow, highlighted by the fact that almost *all* British cytoblastemists, as well as Schwann himself, were up until 1868 ardent *opponents* of spontaneous generation, so that it may be inferred they saw no such implication in a cytoblastemic theory of the origin of cells.[48] Beale, having claimed such a linkage, was an exception on the British scene. But he was the exception that proves the rule; his extreme vitalism had always made his views far from the mainstream.[49]

Why would spontaneous generation opponents feel Bennett's ideas to be such a threat? The physician attending Bennett at the time of his death, Dr. William Cadge, said "Lister was about the worst, and Bennett I suppose [was] about the best teacher Edinburgh ever had."[50] Most of Bennett's obituaries similarly praised his effective teaching ability. One author wrote: "It is no abuse of language to say that Dr. Bennett was a great teacher. Earnest, exact, methodical, practical, he had the power of communicating to others his . . . method of working."[51] And a more recent summary noted, "He differed from many of the great lecturers of his time in that his lectures were fully written out before he gave them. This ensured precision, but his pupils recalled their joy, when he laid aside his manuscript and, *ex tempore,* attacked the opinion of an adversary."[52]

Thus, Bennett's ideas had very wide influence among medical men (just as Sharpey's did with physiologically inclined men of science), because of his powerful position as a respected author of textbooks and teacher in a prestigious medical school for over two and a half decades. This reputation must be balanced against the dismissive pronouncements made against his molecular theory by the 1870s in order to obtain a true picture of the length and breadth of Bennett's influence. The misperception that during the 1850s and 1860s most British scientists perceived Virchow's arguments about cell theory to be relevant to the spontaneous generation debate is also dispelled by this analysis. Bennett's 1868 paper seems to have been uniquely important in connecting the two discourses in British science at that time. Even then, Britain's leading advocate for spontaneous generation, Bastian, preferred to focus on the origin of bacteria and what Bennett would have called histolytic molecules. Yet Virchow's linkage of 1877 has persisted in biology textbooks because the winners write the history books, and Virchow's slogan was successfully adopted.[53] As a result, the pivotal role played in 1868–69 by Bennett's theory has faded into obscurity, as has the consensus of ideas it helped consolidate, leading to a period from 1868 until 1875 or so in which Listerian surgical claims were viewed with great skepticism, and spontaneous generation theories could seem reasonable even in the British context. With Bennett's demise, the term "histological molecules" all but disappeared from medical and biological literature, and with it went a major discourse in which spontaneous generation (at least heterogenesis) could make sense. Indeed, between 1873 and 1875 not only Bennett, but Liebig, Pouchet, Jeffries Wyman, and Robert Grant all died, removing suddenly most of the prominent older figures that had opposed Pasteur and supported spontaneous generation.

Brownian Movement Revisited

Stephen Brush has pointed out that from 1828 to 1870, the physical scientists considered the phenomenon of Brownian movement clearly within their domain, notwithstanding disputes about the precise cause. My analysis has shown, however, that biologists began to adopt this position only in the late 1860s, with Bastian, Huxley, Charles Robin, and Carpenter as early examples.[54] Furthermore, at least a significant minority of widely respected figures in biology and medicine, such as Addison, Bennett, and Owen, as well as medical factions represented by the *Lancet* and the *British Medical Journal,* did not accept this usurping of the "molecules" themselves by the physicists. These researchers continued conceptualizing "molecules" as directly related to spontaneous generation. It can also be argued that more "conservative" particles of generation, such as Darwin's gemmules and Beale's bioplasm particles, were also inspired by Brown's ideas. Darwin himself was still trying to imagine the particles that might survive boiling in infusions to be his gemmules, at least as late as 1870.[55] Standard claims that "later in the nineteenth century [i.e. after Brown], the doctrine of organic molecules was to be swallowed up by the cell theory,"[56] or that all microscopists considered connections between Brownian movement and spontaneous generation "fantastic," are thus missing out on a significant chapter in the development of biology.

A piece in the *Monthly Microscopical Journal* of July 1871 shows that at least until that date the opponents of spontaneous generation still feared that the misinterpretation of Brownian movements of lifeless particles was a major source of reports of spontaneous generation.[57] And in June of 1873, Pode and Ray Lankester continued to hammer away at the same theme, suggesting that this could explain away many of the observations of both Bastian and Child.[58]

As we have seen from a closer examination of Huxley's maneuvering, his attempt to make the spontaneous generation camp vanish from histories of "serious" post-1860 biology was a politically informed act. And Bastian's acceptance of the physical claim to Brownian movement, trying to remain a physicalist despite his rejection by the X Club, played into their hands, giving away the last opportunity in which spontaneous generation could have claimed support from Brownian "molecules."

Life Cycles in Infusorial Monads

One of the most effective arguments undermining spontaneous generation during the mid-1870s has been very little noted and discussed in the secondary literature.[59] This was a series of laboratory investigations into the life cycles of "monads" (protozoa) carried out between 1873 and 1875 by the Reverend William Dallinger and his friend Dr. John Drysdale, both of Liverpool. These experiments will be discussed in some detail, but the larger social and cultural context of Dallinger's work first needs some explication.

Raised in a liberal Anglican family, Dallinger had shown talent in mathematics and sciences as a young man.[60] He came across a copy of sermons by John Wesley and was moved by it sufficiently to become a Methodist minister at age nineteen. The Wesleyan movement in general had long been viewed as hostile to science, and Dallinger had to struggle to maintain his interests in both areas. In an attempt to resolve this tension, he joined others in forming the Wesley Scientific Society. This organization attempted to educate young Methodists in the sciences and to propagandize for compatibility between science and religion.

Even earlier, Dallinger was instrumental in the founding of the Christian Evidence Society by liberal Anglicans and Nonconformists in 1870. This group aimed to stem the threat of infidelity and outright atheism growing in Victorian society. Its scientist members, such as Dallinger, hoped to find ways of accommodating new scientific findings within a liberal theological framework, so that large segments of society that accepted such scientific discoveries as evolution would not feel that they must automatically reject religion as a result.[61] By 1878, Dallinger had developed and "espoused a form of theistic evolution and advanced a novel teleology based on Darwin's description of adaptation in animal forms. His evolutionary system represented the expression of the divine will, which saw man as the apex of nature and excluded the evolution of life from inorganic matter."[62]

Clearly, Dallinger's experiments beginning in 1873 were informed by a desire to disprove Bastian's claims. The popularity of scientific evidence for spontaneous generation was exactly the kind of sharp wedge between religion and science that was feared most by liberal churchmen sympathetic to evolution. Further, Methodist writings since early in the century had explicitly opposed spontaneous generation or matter having "generative qualities," as this tended to exclude divine agency.[63]

It should be noted that the experiments appeared in print in the *Monthly Microscopical Journal,* one of the journals edited by Henry Lawson. Lawson had recently converted from being a staunch supporter in print of spontaneous generation, and of Pouchet, Bennett, Owen, and Bastian, to using the editorial columns of his journals to oppose them and support Huxley and Tyndall. Each installment of the series of papers was read before the Royal Microscopical Society and appeared shortly after in the journal. The entire series occupied seven installments spread out from the August 1873 issue through the May 1875 issue.[64]

Opening their first paper on the subject, Dallinger and Drysdale observed that the spontaneous generation experiments were a jumbled mess of contradictory observations about microorganisms. They attempted to break the log-jam by pointing in a new direction, stating:

> The appearance or non-appearance of organic forms in certain infusions placed in sealed flasks or tubes . . . is held to be decisive of their production *de novo* . . .; but, in point of fact, we know *nothing*—absolutely nothing—of the life history of the greater number of the forms produced . . . At least this is inevitable, that before we can be scientifically certain that these lowly forms do or do not originate in non-vital elements, we ought to know their life history; and if this be desirable in the question of Abiogenesis, it must be absolutely essential before we even approach that of Heterogenesis.[65]

They were suggesting that the radical idea of heterogenesis might be disproven if the transformations of forms attributed to that process were in reality lawful changes contained within the life cycles of stable species. This will be discussed further in the next section. Dallinger and Drysdale declared that the a priori likelihood was to "expect not a *de novo* production, but a production from genetic elements."[66]

To pursue this strategy experimentally, however, was difficult, as continuous unbroken observations over hours or even days might be required to follow the complete life cycle of an organism without missing the critical moment at which it might transform into a different phase. The two men shared an interest in microscopy, and they decided to work together so that by regularly relieving one another at the microscope they could keep up the unbroken vigil. They also needed to develop a procedure for keeping the drop of fluid from drying out or overheating during long continuous observation. This was particularly difficult when observations were carried out at very high powers. Though "monads" (protozoa) are much larger than bacteria, higher powers were necessary to see some of the tiny reproductive spores released by the organisms. Successfully solving these problems was a major breakthrough in microscopic technique, and the two men were lauded on all sides for the resulting significant contribution to understanding life cycles in protozoa. Dallinger, in fact, went on to become president of the Royal Microscopical Society.[67]

Because they conducted their observations at high powers, up to as much as 2,500x, Dallinger and Drysdale reported seeing granules of an exceedingly tiny size released by the monads they studied. These grew into new monads when observed over several days. Their tiny size meant they could have been present in some spontaneous generation experiments but remained invisible at even 800 or 1,000x, so that monads could have appeared in fluids mistakenly judged to be lifeless.

Dallinger and Drysdale were further interested in whether these granules might be more heat-resistant than the adult monads. This would have even greater potential to discredit spontaneous generation as the cause of monads appearing in heated solutions. They dried slides with the granules on them and then heated them in a dry oven that was raised to 121°C. The slide was slowly cooled and rewetted with distilled water, and in some cases the spores were seen to germinate and produce new monads. Eventually they observed development of monads in an infusion raised to 127°C, "suggesting that the sporule is uninjured at a temperature considerably above that which is wholly destructive of the adult."[68] They then directly challenged Bastian's claim that "even if Bacteria do multiply by means of invisible gemmules, as well as by the known process of fission, such invisible particles possess no higher power of resisting . . . heat than the parent Bacteria themselves possess." The monad evidence showed, they said, that "[t]his *may* be true of Bacteria, but it certainly remains to be proved; while its inapplicability to *all* sporules is apparent."[69]

Early in 1876 Dallinger contacted Darwin to compare notes on microscopic work at high powers.[70] In his reply, Darwin brought up the earlier papers: "I have read all your and Dr. Drysdale's papers, and they seem to me to possess higher value than anything which has been published on such subject, though I am too ignorant to have any right to express such an opinion. But I have a full right to say that they are extremely interesting."[71]

At about the same time, Henry Lawson put Dallinger in contact with Tyndall.[72] When Tyndall had read the papers, he wrote Dallinger to say that he was especially impressed with their proof that the differing stages in monad life cycles were yet regularly repeating and therefore *not* examples of heterogenesis: "With regard to . . . your researches, . . . I hardly ever read anything that interested me more . . . I wish you health and opportunity to pursue these truly admirable researches. If you can do for the Bacteria what you have done for the Monads it assuredly will be a great achievement. But whether you do it or not, that which you have done gives you a permanent name in science."[73]

No such work was done by Dallinger, but this shows how clearly focused Tyndall had become on trying to find (or encourage) work that would prove that bacteria also have life cycles, especially in which one stage was a resistant spore.[74] When Tyndall was handed Cohn's *Beiträge zur Biologie der Pflanzen* a few months later, containing Cohn's account of spores in *Bacillus subtilis* and Koch's report of a spore stage in the life cycle of *B. anthracis*, his mind was clearly prepared to seize on those findings. And between Dallinger's and Tyndall's experiments, Darwin, after over a decade of waffling, finally began to come to a position in which he could accept that experimental evidence was clearly against spontaneous generation.[75] These final experiments of Tyndall will be taken up in chapter 7.

Pleomorphist Theories of Bacteria

Since at least the writings of Karl von Nägeli in the 1850s, germ theories of disease and theories of spontaneous generation have been bound up in complex ways with notions of bacterial pleomorphism. This view, particularly prevalent in Britain from the late 1860s to at least the 1880s, held that because they are such primitive organisms, there are no distinct and stable bacterial species; rather, that bacteria transform through different stages, even from pathogenic to nonpathogenic forms and vice versa, under differing environmental conditions. Nägeli was among the first to publish this view of bacteria (which he called schizophytes), and almost from the first he argued, in the words of one historian, that "[i]t is these that are produced everywhere and at all times, by spontaneous generation; these are the youngest, the most recent graduates, as it were, from the inorganic world. Their characteristics, Nägeli felt, were just what might be expected from such a youthful group: their multiple forms, so similar to each other, and so easily transforming into each other, make varieties, species and genera impossible to fix."[76]

This widespread belief in the primitive, homogeneous, and inconstant nature of bacteria was an extremely important reason why so many believed that bacteria could be spontaneously generated, even if more complex organisms such as algae and protozoa could not. It was only the work of Ferdinand Cohn, a Breslau botanist who finally constructed a taxonomy of bacterial species beginning in 1872 which convinced researchers that bacteria could be treated as constant species.[77]

In Britain, T. H. Huxley, E. Ray Lankester, and Joseph Lister were all strong advocates of pleomorphist views in the early 1870s.[78] In the summer of 1870, while Huxley was preparing his attack on Bastian for the Liverpool meeting, he and Hooker exchanged a series of letters in which Huxley reported observing the interconversion of bacteria, yeast (torulæ), penicillium, and even a type of algae (Ulvella). Huxley reported, "I am working just now at the yeast plant with high powers—and if I have not got the development of *Bacterium* straight out of it, call me horse."[79]

And a few days later he continued:

> If you make an infusion of hay . . . and examine it, within 24 hours you find it full of the free particles (=*Monas termo* and *Bacterium termo*), and the pellicle which forms at the top of the infusion in two days is the *Ulvella* form of the same thing . . . The *Ulvella* particles may elongate into rods and become *Bacteria* or multiply and form long chains as . . . *Leptothrix, Leptomites,* or whatever other name you may choose. This is to be seen very easily . . . Now I can in no wise distinguish many of these *Leptothrix* forms from the finer portion of the mycelium of

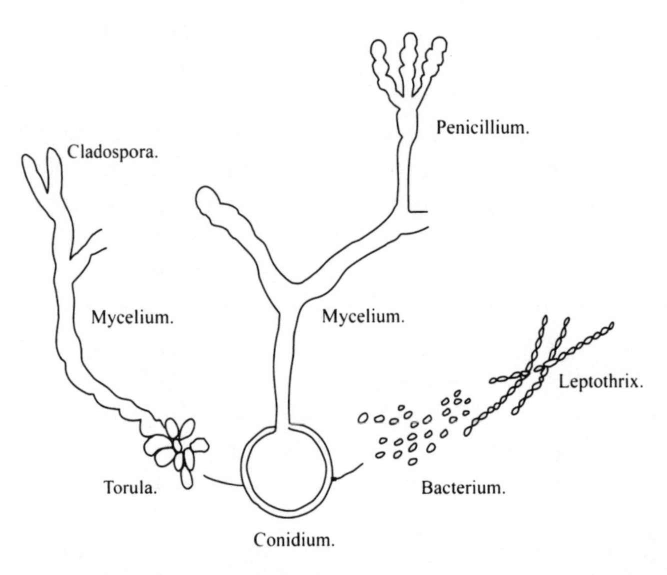

T. H. Huxley's diagram of the hypothesized pleomorphic relations between yeast (torula), molds (*Penicillium* and *Cladospora*), and bacteria. From his paper of 1870 in *Quarterly Journal of Microscopical Science*, "Penicillium, Torula and Bacterium."

> *Penicillium* . . . and I believe, though I cannot prove it yet, that they grow into that mycelium.[80]

Both men freely speculated upon the evolutionary implications of such transformations. Hooker, for instance, wrote that perhaps this offered the clue to how the tissues of higher plants differentiate from the cambium layer:

> if *Ulvella* really gives rise to *Penicillium,* so may it to these large, ill-defined spores . . . which I thought so like *Penicillium* spores . . . I imagine its retrogression to *Ulvella.* Thus *Penicillium*'s regression to *Leptothrix* filaments. After all, why should we not have here an epitome of what goes on from Cambium, which must consist of [Bacterium] *Termos* . . . under circumstances that eventuate in the various tissues that the plant is composed of, from the simplest cell to the most complex pollen grain . . . You have made an immense leap in the association of forms, and I cannot but suppose you approach the final solution.[81]

In other words, much as Oken would have put it, the multicellular plant (and its cambium layer) are just colonies of simpler units (the *Bacterium termo* particles). In his next letter Hooker continued rhapsodizing on the importance of the pleomorphic discoveries.[82]

Huxley's support of pleomorphism was consistent with his view of *Bathybius* (prior to 1875) as undifferentiated primordial living protoplasm, which I have shown was widely viewed as implying that Darwinian evolution must also be based on abiogenesis, or at least heterogenesis.[83] Bastian declared this context central to his argument for heterogenesis in the opening pages of his magnum opus *The Beginnings of Life:*

> Again, we find that the comparatively low forms of life in which all these developmental transitions are embodied, instead of being almost unchangeable . . . are variable in the highest degree. They pass through the most diverse and astounding transformations, and . . . are often seen to be derived from matrices wholly unlike themselves. In fact, these lower forms of life—corresponding pretty closely with the Protista of Prof. Haeckel—form an enormous and ever-growing plexus of vegetal and animal organisms, amongst which transitions from the one to the other mode of growth take place with the greatest ease and frequency. Here Heterogenesis is constantly encountered and variabil-

> ity reigns supreme, so that those assemblages of definitely recurring individuals, answering to what we call "species," are not to be found amongst them.[84]

This view was also a crucial reason why Bastian felt that new developments in the study of heredity, including cytological studies on mitosis, did not invalidate his work. More stable types could eventually be produced from the process, but "until such assemblages of repeating individuals make their appearance—that is, until Homogenesis becomes the rule—the 'laws of heredity' can scarcely be said to come into operation."[85] Bastian never changed his position on this point, only elaborated it since his opponents talked past it. Bastian continued to argue this when, after 1900, he was accused of being still more out of date after the rediscovery of Mendel's work. To anyone who accepted the observations of variability as proof of the easy transformability of one species into another, it would be clear that none of the recent discoveries had any applicability to this most primitive realm of the forms transitional between nonliving and living. Bastian's very small following after 1900 shows that few of the new generation of biologists, faced with this dilemma, were willing to accept the transformability as fact. By this time Koch's doctrine of monomorphism of bacterial types was at its height, as was his fame for isolating causative disease microbes.

In 1880, Bastian wrote of the ability of low organisms to change rapidly from the animal to the vegetable state and vice versa: "we are compelled to believe that such passages from the one mode of molecular composition and activity to the other, may be determined without any great difficulty by internal chemico-nutritive changes."[86] And he argued that such "changes of state" were analogous to changes of form among simpler kinds of matter: "It is certain, as Prof. Graham showed, that one and the same saline substance may exist with its molecules now in the crystalloid and now in the colloidal mode of aggregation, according to the different influences under which it has been produced . . . [I]t is also known that certain typical colloids may, under some conditions, be converted into crystalloids."[87]

Bastian went on to say that the allotropic forms among which carbon, phosphorus, and sulfur were capable of metamorphosing were further analogous cases. Thus, he argued, the animal and vegetal forms and their interconversion in the lower organisms are "strictly comparable to" these chemical transformations, so that animal and vegetal "may be regarded as different allotropic states of Living Matter."[88]

Because pleomorphic transformation was seen as supporting spontaneous generation, William Roberts took the trouble in 1874 to refute Huxley's earlier claim that bacteria could develop from torulæ or vice versa.[89] In this context, Huxley's retreat from pleomorphism over the next couple of years was clearly related to his recognition of the effectiveness with which Bastian was using the concept. Still, although Huxley and Lister had retreated from any pleomorphist statements by 1876 or so, Ray Lankester continued to be its most prominent advocate and to oppose Cohn and later Robert Koch over the issue of bacterial species well into the 1880s.[90] Koch from the first linked his monomorphism to his claim that specific bacterial species must cause each specific infectious disease. Furthermore, both Cohn and Koch argued against spontaneous generation as part and parcel of their project of creating taxonomies. Thus, the existence of vocal opposition to Cohn and Koch's project was an important contributing factor in Britain to an intellectual climate in which spontaneous generation could be viewed with sympathy, despite the irony of the fact that Huxley, Lankester, and Lister all believed that current-day spontaneous generation was not proven, and that Bastian was wrong.

Transformism of microbial forms (grading into heterogenesis in the logical extreme) was seen by many as analogous to transformism of species over longer time periods. And Nägeli was not the only writer who saw support for Darwinian evolution as linked with support for pleomorphist doctrine. Pasteur himself, who opposed both doctrines to some extent, argued that they were connected, in a rare direct reference to Darwin:

> Thus we shall be led to believe, in all good faith, that we have under our eyes examples of the changes of spores of fungoid growth into cells of ferment, or proofs of the conversion of *bacteria* . . . into the same cells. Causes of error of this nature have induced some German naturalists [e.g., Hallier] to believe that they have succeeded in proving . . . that a number of fungoid growths . . . may become transformed into yeast . . . We have combated them since the year 1861. Since that period they have lost rather than gained ground abroad, in spite of the

> growing favour bestowed on the Darwinian system. One of the mycologists who enjoy the most legitimate authority beyond the Rhine, M. de Bary, has arrived, as we have, at absolutely negative results.[91]

Huxley's paper of autumn 1870, "Penicillium, Torula and Bacteria," suggested that those three forms were merely pleomorphic phases of the same organism (see figure). Thus, once again, a paper of Huxley's was widely perceived to suggest a connection among Darwinian evolution, pleomorphism, and perhaps even heterogenesis of microorganisms. The abandonment of pleomorphist views by Huxley and Lister went far to weaken this pillar of support for belief in spontaneous generation in Britain at just about the time of the final major confrontation between Bastian and Tyndall in late 1876 and early 1877.

6

Germ Theories and the British Medical Community

After Bastian felt his access to the Royal Society had become limited and saw the pages of the *Quarterly Journal of Microscopical Science* and the *Monthly Microscopical Journal* become hostile to him, he turned to the Pathological Society of London, the Clinical Society, the *Lancet,* and the *British Medical Journal.* This chapter explains why this was a reasonable thing to do. To understand this move, we need to understand more about the relationship between the doctrine of spontaneous generation and the germ theory of disease.[1]

In Britain by the mid-1860s medical views on the nature of contagion were complex and varied. Pasteur had promulgated the idea that living microorganisms or "germs" of them (analogous to seeds) were the actual agents of contagion. The early response of British doctors was lukewarm: they did not immediately cast out established Liebigian models of "zymosis" (disease caused by a process of catalysis from a rapidly multiplying chemical poison). This model had spread widely because so many prominent British scientists and medical researchers were students of Liebig, including Thomas Graham, Robert Angus Smith, Henry Bence Jones, and J. L. W. Thudichum. Although Pasteur viewed his own model as totally incompatible with Liebig's, British doctors did not assume this; instead they tried to incorporate the new findings about bacteria into existing zymotic explanations of disease. For example, John Simon, medical officer to the Privy Council, set Thudichum to work on the chemical poison and John Burdon Sanderson to work on bacteria in 1867. There was room for a variety of views on the role for microorganisms in the disease process, and one's position on spontaneous generation similarly could be accommodated to any of a variety of "germ theories of disease," or to outright opposition to any causative role for germs.[2]

Many British doctors with laboratory training, for instance, were arguing by the 1860s that gases from sewers, even bad-smelling ones, were *not* in all likelihood the agents of disease.[3] These were not the kind of simplistic "anticontagionist believers in miasma" as the medical or sanitarian opponents of the germ theory have often been portrayed.[4] Yet while many, such as William Farr, Benjamin Ward Richardson, Henry Bence Jones, or John Hughes Bennett, believed in some specific particulate contagion, they did not unreservedly accept Pasteur's claim that living "germs" were the contagia. Farr and (by 1867 at the latest) Bennett, indeed, believed that the microbes were spontaneously generated. The physiologist William Carpenter strongly opposed spontaneous generation, especially after the late 1840s when he experienced a renewal of religious feelings, yet he also forcefully opposed germs as sole causative disease agents. When some, such as John Burdon Sanderson, Joseph Lister, and Lionel Beale, began to argue for a "germ theory of disease," their theories differed greatly from one another, and from the versions being promulgated by nonmedical advocates such as Pasteur and Tyndall.[5] Thus, as this chapter will show, the very multiplicity of germ theories of disease made accommodation with spontaneous generation possible for longer than has been supposed. Even if one bluntly opposed a causal role for germs in any form, as Bastian did, the nonunified front presented by these multiple, somewhat contradictory germ theories made them an easier target. To many, in fact, it seemed reasonable to suppose that bacteria could be produced by spontaneous generation, as by-products of the disease process. I will return to this point near the end of this chapter.

The Cattle Plague of 1865–66 and Germ Theories

An epidemic of Siberian Rinderpest began among British cattle in June of 1865. By October over 11,000 cases had been reported, and a Royal Commission had been appointed to report to Parliament, particularly because the epidemic led to embargos on British beef in France and Belgium, damaging British trade noticeably.[6]

Edwin Lankester, M.D., a well-known microscopist and editor of the prominent *Quarterly Journal of Microscopical Science*,[7] commented on the epidemic in his address of 9 October 1865, as president of the Public Health Department of the Social Science Association.[8] His address serves as an authoritative statement surveying medical opinion of the moment. There were two great schools of thought among sanitary reformers, he tells us,

> the disciples of one of which maintain that the great mass of zymotic diseases are produced by special poisons and are called contagionists. Whilst the followers of the other school do not believe in special poisons, but maintain that certain general conditions of sanitary neglect and dirt are alone necessary to produce the group of zymotic diseases, and they are called anticontagionists. Now I believe the extreme views of each school are wrong; . . . for an officer of health to suppose that cleansing, and draining, and washing, would arrest the progress of small-pox in a house full of unvaccinated persons, would be an utter absurdity; whilst the placing a cordon around an ill-ventilated and badly warmed house, expecting to keep off bronchitis and pneumonia, whilst the temperature is 12° below freezing point, would be equally absurd.[9]

Note that Lankester identified "contagionists" at this time as supporters of the *zymotic* disease model, not the germ theory model. And yet his description of the nature of the "poison" included numerous different characteristics. First, he stated, "It seems demonstrated that there arise, during the decomposition of vegetable and animal matters, certain organic molecules, which being taken into the system, will produce certain definite changes in the system constituting well-known forms of disease. Thus, ague and kindred fevers . . . and paroxysmal fevers . . . [are produced] by poisons formed during the decomposition of vegetable matter."[10]

He distinguished, however, between those contagious diseases and many much more common ones including smallpox, scarlet fever, measles, whooping cough, typhus fever, typhoid fever, and cholera. Regarding the nature of the poisons responsible for these, Lankester stated that "we know but little; but recent researches with the microscope lead us to hope that we are not far distant from the time when, at least, the form of these poisons will be made visible to the human eye."[11]

But he was referring *neither* to Pasteur's microbial germs *nor* to the granules of "germinal matter" produced from diseased decomposing tissues that Beale had suggested as contagia. Instead, Lankester argued that it was the *white blood cells* that accumulate in the pus of smallpox pustules, for example, that conveyed the disease. These he called "the germs of the poison" and "poison-cells . . . or germinal elements"; furthermore, he said this explained how they could be conveyed in dried vaccine lymph (in the case of cow-pox), "in linen, cotton, and woolen fabrics, how they can be conveyed in letters and newspapers . . . and only need the awakening influence of a little moisture to summon them to awaken and live anew."[12] And regarding scarlet fever, typhus, cholera, and the others of this group, he continued:

> reasoning from analogy. . ., we are driven to the conclusion that these diseases depend on a cause similar to that of small-pox, and that the real form of the poison is the charged white cell of the blood. It is, however, interesting to notice some varieties in the habits of these poisons. Thus we are not aware that the poisons of small-pox, measles, scarlet fever, or typhus, are conveyed by any means but through air; whilst there is every reason to believe that the poisons of typhoid fever and cholera are conveyed by the agency of water.[13]

This gives some flavor of the varieties of understanding and use of "germ," as well as of the continuum doctors easily constructed within "zymosis" to accommodate living agents as well as Liebigian chemical "poisons" that started disease because of their ability to transmit their chemical instability in a cascading fashion in the tissues of an infected organism.

When the Royal Commission interviewed John Simon, the medical officer of the Privy Council, he echoed most of Lankester's opinions, taking a contagionist line on the nature of the cattle plague. But again, it should be emphasized, "contagionist" did not necessarily mean a simple embracing of the germ theory as it is retrospectively viewed today. The "contagionists" had a complex mix of views on what the nature of the contagion actually was. Robert Angus Smith, a Manchester chemist trained by Liebig; Burdon Sanderson; and Beale carried out laboratory investigations on the disease and autopsies on the dead animals. Burdon Sanderson reported that the blood of an infected animal contained some discrete agent capable of infecting a healthy animal. He first called the infectious particles a "poison." Only by 1869 did he begin referring to them as "microzymes," a term he used more or less synonymously with what others called "Bacteria" through about 1874. Beale, on the other hand, in line with his earlier theory reported that the agent of infection was particles of living "germinal matter" that multiplied in the blood of an animal from a few particles to numbers great enough to clog up circulation in the capillaries.

Angus Smith had, prior to the cattle plague, been somewhere between a miasmatist and a believer in purely chemical zymotic poisons.[14] His experiments on the effects of disinfection and ventilation on cattle plague began to make him think seriously about atmospheric germs (in Pasteur's sense).[15] But it was not until he had carried out microscopic observations with his friend J. B. Dancer in 1867–68 on atmospheric dust collected from the air of Manchester that he began to focus intensely on living germs as the main agents of contagion.[16] Thus, in the 1866 Reports of the Commission on the Cattle Plague, neither Smith nor Burdon Sanderson nor Beale expressed the germ theory as we know it today. Histories that describe their work on cattle plague as the turning point in acceptance of "the germ theory" in Britain have disguised the ambiguity and heterogeneity of medical opinion and terminology that continued through the mid-1870s. It is precisely this ambiguity, I argue, that created a theoretical space for spontaneous generation to flourish for a decade or so in Britain, beginning in 1868.

Similarly, Lister's system of antiseptic surgery (wounds protected from exposure to some particulate airborne matter showed greatly reduced rates of infection) did not cause any sudden turn of opinions toward the germ theories in Britain. For one thing, Lister's early rhetoric did not tie his recommended changes in practice inseparably to a Pasteurian germ theory. As late as 1871, in an address to the British Medical Association annual meeting, Lister said, "I do not ask you to believe that the septic particles are organisms . . . [I]f anyone . . . prefer[s] to believe that the septic particles are not alive, and to regard the vibrios invariably present in putrefying pus . . . as mere accidental concomitants, . . . or the results, not the causes of the change, with such an one I, as a practical surgeon, do not wish to quarrel. Nor do I enter upon the question whether spontaneous generation can take place at the present day upon the surface of our globe."[17]

It has also been shown that Lister's system itself evolved so much from 1867 through the 1870s, in response both to criticisms and to trying out new variations, that referring to "*the* method of antiseptic surgery" during the period of the spontaneous generation controversies is as much of an ahistorical reconstruction as referring to a single unified germ theory at this time. Lister himself in the late 1870s and early 1880s began such a retrospective reconstruction for his own rhetorical purposes.[18]

Perhaps even more important, however, was that John Tyndall became the most prominent lightning rod for medical response to the germ theory. And Tyndall's personal brashness and excessively simplistic version of the germ theory both provoked anger as well as skepticism from doctors.

Tyndall, the Germ Theory, and the Medical Community

On Friday 21 January 1870 Tyndall gave one of his celebrated lectures at the Royal Institution, titled "Dust and Disease." There he stated that the organic nature of dust suggested strongly that Pasteur was right, and that airborne germs were probably the real source of disease, especially contagious disease. The origin of Tyndall's interest in the germ theory requires some explanation, since none of his biologist friends in the X Club had taken any position on the subject beforehand. Barton points out that the X Club contained no active medical practitioners among its members.[19] Tyndall, like all Victorian men of science, had commented on the need for science to seek a way to fight the Cattle Plague of 1865–66.[20] Perhaps it was Bence Jones, Tyndall's friend who had served on the Government Commission on the Cattle Plague, who mentioned to him Lister's development of antiseptic surgery, first announced in 1867. Bence Jones had expressed respect for Pasteur's work, but as recently as 1867 had nevertheless remained unwilling to declare that spontaneous generation had been disproven as a source of disease organisms.[21] By 1870, just after the appearance of a paper about Pasteur's work by Tyndall, Bence Jones wrote to congratulate Tyndall, but to express continued uncertainty.[22]

The chemist Edward Frankland had become a convert to the germ theory in 1866. It has been argued that it was he who, in regular contact with Tyndall at meetings of the X Club, supplied the impetus for the physicist to see atmospheric dust particles in the way Frankland saw water-borne particles: as agents of disease.[23] According to Tyndall himself, however, his interest was provoked only after he was surprised to discover, on 5 October 1868, that the dust suspended in the air was largely organic in nature: "I was by no means prepared for this result; having previously thought that the dust of our air was, in great part, inorganic and non-combustible."[24] Even then, the analogy with Lister's results does not seem to have solidified into a certainty in his mind until the following summer or autumn, when Tyndall was vacationing at Bel Alp in Switzerland. In September 1869, after bathing in a rocky Alpine pool, he slipped and deeply wounded his lower leg on a sharp granite outcrop. The wound was at first showing signs of healing cleanly with no inflammation. A few days later, after he began to walk on it prematurely, an infection did set in, spread to his foot, and became so severe as to be life threatening. Tyndall's interest in Lister's technique probably was stimulated by this experience: he began to suspect that bacteria from the air, getting into his open wound, were the cause of the immense abscess that developed.[25] Very likely, *all* these were involved in awakening Tyndall's interest for dust.

By late 1869 Tyndall imagined that dust particles must act in the disease process as carriers of invisible microbial germs. Thus, dust was not only an organic product, but also an infectious agent. Bastian, too, considered that Pasteur had proved dust to be involved, but Bastian claimed this proved only that dust was a necessary ingredient for spontaneous generation. The final catalyst for Tyndall was his meeting, only a month or so after recovering from his illness, with William Budd, a germ theorist physician. Tyndall took an immediate liking to Budd, and the doctor may have expressed his opinion at this meeting on the airborne bacterial origin of Tyndall's recent erysipelas.[26] In almost all of his future lectures and papers against spontaneous generation, beginning with the very first in January 1870, Tyndall effusively praised Budd, crediting him with first standing up for the germ theory in Britain and thus being more far-sighted than most British medical men. When both the *Lancet* and the *British Medical Journal* ridiculed Tyndall for presuming to tell doctors what caused disease, Tyndall defended not only his pride but his certainty in the methods and theories of Budd and Lister.[27]

The *Lancet* reviewer was critical in particular of Tyndall's "hysterical tone," which he likened to that of *Vestiges of the Natural History of Creation* in terms of its sensationalized reception in the press. The weekly *Scientific Opinion,* though more reserved in tone than the medical journals, also treated Tyndall's conclusions with skepticism: "We do not go so far as to say that spontaneous generation is demonstrated; but we must confess that it appears to us that both Dr. Tyndall's and M. Pasteur's experiments are, as Dr. Bastian says, of a very ambiguous character. We are further of opinion that the problem is one . . . to be wrought out by the biologist, and not by the physicist; and we think it somewhat unjust of Dr. Tyndall to ignore the fine memoirs of both English and foreign biologists."[28]

Despite his new interests, Tyndall did not immediately reject the views of Bennett or Bastian. He repeated his support for Pasteur's germ theory in his initial letter to the *Times* of 7 April 1870, and only when Bastian's reply seemed to him to insult his many years of experience as a laboratory scientist did Tyndall begin to treat Bastian as though he doubted the younger man's scientific reasoning. And it was only after Huxley's pronouncements during the summer of 1870 that Bastian's skills as a laboratory experimenter were not up to par that Tyndall began to question what had been up to that time the strongest base of Bastian's scientific reputation. In this mood, Tyndall must have had little patience with a letter that arrived from Felix Pouchet, dated 8 June 1870. The French heterogenist politely mocked Manchester microscopist J. B. Dancer's claims about dust containing germs (and by implication Tyndall's work, which, like Angus Smith's, leaned on Dancer's): "I will make my renunciation if . . . you will let me see the spores that Mr. Dancer says he has discovered in such great numbers in the air. It would be easy to send them to me, between two sheets of tissue in a letter . . . But in spite of the great precision of my methods, I have never collected anything similar. On seeing these things, I am certain, there is not a heterogenist who would not renounce his illusions."[29]

Similarly, once Bastian had demonstrated what the X Club considered brashness and independence from their advice to "tone it down," he became a target for Huxley and especially for Tyndall. He became a convenient focus for many of Tyndall's long-standing gripes: the medical press and "quackery," the sad state of British science education as compared with the Continent, proper intellectual requirements and behavioral restraint for first-rate scientists, and so on. It is ironic that such organs of Victorian science as the *Journal of the Quekett Microscopical Club* had for the previous two years declared Bastian one of the rising stars in microscopical skill and lauded him for his elegant and persuasive reasoning on the origin of life, while in the same breath touting microscopy as guaranteed to develop "the Moral Qualities," most of all patience.[30] Yet the centerpiece of Tyndall's attack on Bastian was the claim that the young pathologist was not only experimentally incompetent, but also morally deficient, with an impatient desire for public acclaim.

Furthermore, Tyndall's long-standing campaign against organized religion came to a head during this period over an episode in which clergy and their prayers were publicly acknowledged for curing the Prince of Wales of typhoid fever, whereas the medical treatment he received went unnoticed.[31] To add insult to injury, the debacle at the 1874 Norwich meeting of the British Medical Association provided a big boost to the anti-vivisection movement in the eyes of the British public, which displeased Tyndall. Here protestors had interrupted a demonstration on live dogs, leading many onlookers to sympathize with the anti-vivisection cause.[32] During the same period, 1871–72, Tyndall came to see not only religion and anti-vivisection, but most of the medical profession itself, especially its journals, as obstructing scientific progress in medicine because of their resistance to the germ theory.

In response, Tyndall was opposed by a wide segment of the medical community, not only by Bastian, and many repeated the charge that the physicist was out of his depth when speculating on biological problems. These critics included very respected names such as Charles Murchison, *Lancet* editor James Wakley, Henry Lawson, and Lionel Beale. Beale could be quite offensive in public, and not less so in private. Though critical of Bastian, he was even more critical of Huxley and Tyndall. In a letter to Burdon Sanderson, he mocked Tyndall's showmanship in his well-known public lectures at the Royal Institution: "Tyndall may dance on the tight rope & put the blue sky into his snuff box, but surely we who work at living things need not suggest that things *may differ* indefinitely as regards particulars which lie *beyond* the reach of microscopic investigation."[33]

J. L. W. Thudichum, M.D., a student of Liebig and an influential British physiological chemist,[34] had noted in 1869 how unkindly the medical profession took to members of other specialties attempting to foist simplistic germ theories of disease upon truly scientific medicine. He was referring to the cholera fungus theory of Hallier:

> I ask you, do you think it likely that a botanist . . . with some bottles of cholera excretion . . . could at one stroke, with one glance of his eye,

> discover the cholera contagion under the microscope? I ask you as microscopists whether that is probable? . . . If it were possible that a botanical professor could make such a discovery, and it could be well established, I as a medical man, would to-day abdicate my functions. I would at once retire from the medical profession . . . I would say that I had been a very great impostor to mankind; that I had pretended to possess a certain method of gaining knowledge which I did not possess; that I had been pretending to inquire into some of the deepest problems, and failed to find anything, while a botanical professor had been capable, by a simple scrutiny with the aid of the microscope, to discover that which I had been searching for with all my might for years and years . . . I have spoken to many practical men, and all of them have shaken their head at the fungus theory, and none of them have allowed it to be of such a nature as to bespeak their confidence. Therefore we may dismiss it, and we will continue our own path and our own studies, from which we hope to establish better results.[35]

In 1867 John Simon had given Thudichum "the task of finding 'the very chemical formula' of the morbid process of a zymotic disease, arguing that, 'failing this, we have no final pathology.' By contrast, studies of the germ . . . responsible for a disease . . . were in Simon's view likely to produce only 'generic' or second-order knowledge."[36] And in an 1871 medical microscopy text the balance was still declared to be against the germ theory. Bacteria are sometimes seen in freshly voided urine, noted Joseph Richardson, M.D.; and he said that their occurrence there was difficult to account for unless one accepted the spontaneous generation theory of Pouchet, Bastian, and others.[37] Thus, Richardson concluded that it was premature to declare microbes as causes of zymotic disease, particularly since the weight of so much authoritative opinion was overwhelmingly against the germ theory. Further, he added, "should the pregnant doctrine of Heterogeny . . . be at some time fully demonstrated, it must largely modify any view of the parasitic origin of diseases."[38]

Dr. William Gull, distinguished physician to the Prince of Wales, referred to Tyndall in particular when he stated in his Harveian Oration of 1870 to the College of Physicians that physicians had lately been blamed for being obtuse in their skepticism over germ theory. He reminded his listeners that the theory was not new and that a cholera-fungus had been announced in Britain in 1849, only to turn out to be contaminants that were known harmless fungi. Tyndall had shown that dust consisted of chiefly organic particles, Gull continued. But Tyndall had done nothing whatsoever to clarify whether that organic matter was amorphous material or germs. Thus, he concluded, medicine knew nothing more from Tyndall's researches about the propagation of contagious disease than it ever had.[39]

In December 1870, a *British Medical Journal* editorial, probably penned by Ernest Hart, asked for "A Word with Professor Tyndall." The writer noted that "in some former utterances Dr. Tyndall had astonished men 'mainly occupied with observation'—a class for whom he expresses some contempt—by employing, in public, words tending to confound atmospheric particles with organic germs." And, as if that weren't enough,

> He [Tyndall] now . . . asks the President of the British Association [Huxley], "whom untoward circumstances have made a biologist," but who was intended by Nature for the higher gradational development of a physicist, to "excuse him to his brethren" in the inferior class, when he descends from the seat in the skies which he for the moment shares with Professor Tyndall, and to explain to them that they form an inadequate estimate of the distance between the microscopic and the molecular limit, and that they employ a phraseology which is calculated to mislead. The particular person to whom Professor Huxley is requested . . . to convey this information, . . . [t]he horrid example selected, is apparently Dr. Charlton Bastian, . . . and he is at this moment, we believe, in a state of alarming rebellion, and displays something more than Galilean obstinacy.[40]

The correspondent emphasized that it was Tyndall who was ignorant of the terminology at issue, since he seemed unaware of the histological use of such terms as "structure" or "molecules" in a way quite different from his own.[41] Furthermore, the *British Medical Journal* writer found Tyndall's arrogant tone so offensive that he offered Tyndall this advice:

> It has been said of a well-known political controversialist, that his tone towards those who differed with him was so arrogant that "it would have been offensive in the Almighty towards a beetle." We hope that Professor Tyndall will not object to the saying because it is profane. We trust our readers will excuse it, because it is apt . . . This will perhaps suggest that his tone might be modified with advantage. We are quite sure that science will lose nothing by the change; and that his excur-

> sions from the territory of physics beyond "the utmost bounds of matter on the infinite shores of the unknown" will be none the less profitable.[42]

Clearly, there was no love lost between Tyndall and the medical profession. Many doctors were not impressed by Tyndall's rhetoric about the higher status of physics and chemistry, since they believed that their science had made similar progress in the previous fifty years. Indeed, Richard Quain, M.D., a pathologist and professor of anatomy at University College London, in his 1869 presidential address to the Pathological Society, of which Bastian was one of the most prominent members, went so far as to claim, "Comparing our knowledge then, I speak more especially of morbid anatomy, . . . I do not hesitate to say that the number, the value, and the accuracy of its facts and observations will bear favorable comparison with like progress in any, as they are called, of the more exact sciences."[43]

During 1871–72 the well-known physiological writer G. H. Lewes became a convert to the germ theory of disease after visiting the German laboratory of Friedrich von Recklinghausen. Given this news and an encouraging letter from Pasteur on 3 October 1871, Tyndall felt sufficiently bolstered to issue a second edition of his "Dust and Disease" lecture, including a reply to those who criticized him for intruding into medicine and biology.[44] "'The germ theory of disease,' it has been said, 'appertains to the biologist and the physician.' Where, I would ask in reply, is the biologist or physician, whose researches, in connection with this subject, could for one instant be compared to those of the chemist Pasteur? It is not the philosophic members of the medical profession who are dull to the reception of truth not originated within the pale of the profession itself."[45]

Tyndall again cited William Budd as the model of the enlightened medical man, contrasting him to accomplished figures like Dr. Charles Murchison, who ascribed "epidemic diseases to 'deleterious media' which arise spontaneously in crowded hospitals and ill-smelling drains," and according to whom "the contagia of epidemic disease are formed *de novo* in a putrescent atmosphere."[46] But it was Bastian in particular that Tyndall singled out for condemnation, at first in private letters. And he now began to describe Bastian in the same way he spoke about religion. He likened Bastian's ideas to religious ones because of

Bastian's appeal to the gullible public, and because those ideas had the same grave consequences in loss of life from disease that Tyndall blamed on religious resistance to a scientific understanding of infection.

Tyndall's chances for a positive response from the medical profession were not likely to improve, given his continued haughty attitude toward them in his public pronouncements.[47] Murchison, meanwhile, achieved great acclaim in the fall of 1873 after tracing a typhoid outbreak in the heart of London's professional community to a "poison" (not a germ) spread in contaminated milk, and this boosted the stock of the anti-germ theorists considerably in the public eye.[48] Yet in November 1874 Tyndall wrote a letter to the *Times*, again extolling the virtues of Budd's theory of typhoid spread by microbes.[49] The medical profession was skeptical about Budd's simplistic formula. Doctors were firmly convinced that diseases, especially enteric ones such as typhoid or cholera, involved many factors, and that contamination of drinking water by sewage or sewer gases alone could cause them. Years after Tyndall was publicly considered to have vanquished Bastian, this consensus continued to be widespread in the English-speaking medical community,[50] so that confusion existed as to exactly how the microorganisms and noxious poisons were related. Still, as widely respected an authority on physiology as William Carpenter expressed certainty that Tyndall's version of the story was ridiculously simplistic:

> Have you seen Tyndall's absurd letters in the Daily News? He out-buds Budd, maintaining that cholera and typhoid can *only* be propagated by the introduction of their germs into the alimentary canal; so that if a man's water supply be pure, and he does not take in the intestinal ejecta of a cholera or typhoid patient with his food or drink, he may live close to an open sewer, or over a choked-up cesspool, or have his house filled with sewer gases, without any danger of taking these diseases! He says I belong to an "antiquated school," because I do not agree with him. His authority with the public is such, that I consider it necessary to show that this is a matter on which he is *not* to be trusted.[51]

The active sanitary reformers in the medical profession, who had worked for decades to obtain government support for large-scale public sewage and water systems, and who were convinced that *that* work was the real cause of declining cholera deaths in Britain by the 1860s and

1870s, were not about to allow Tyndall and his newfangled theory to steal all the credit for such hard-won improvements. Most of those sanitarians had a medical understanding of the culpability of dirt and filth as *chemical* poisons, which, by Liebigian catalysis of fermentation-like processes within the body, led to "zymotic" disease. Unlike Tyndall, few of them saw a conflict between the theories of Liebig and Pasteur. Many of them even felt that it was such a chemical process that *generated* the microorganisms that were now found to be associated with disease. In other words, they suspected that the microorganisms were by-products of the disease process rather than causes of it. In this sense, Bastian's theories, explicitly citing Liebig, were in the mainstream of British medical thought, and Tyndall's were not.[52] As mentioned previously, the father of English sanitarians, Edwin Chadwick, was among those that took an interest in Bastian's work and he asked, via Sharpey, for a personal introduction.[53]

In the summer of 1874 an International Sanitary Conference convened in Vienna. There sanitarian views of diseases, especially cholera, dominated. The views of Max von Pettenkoffer, questioning the central role of germs, were particularly respected and were reported in coverage of the conference by the British medical press.[54] Pettenkoffer cast the old "seed vs. soil" dispute over the role of the infecting substance in new terms, invoking an important role for the soil, literally. Pettenkoffer's theory allowed that bacterial germs might have some role in disease production, but only when they interacted with the soil under certain specific groundwater conditions. Thus, for Pettenkoffer, it was impossible for bacteria deposited directly into a city's water supply from sick persons to produce the actual poison that caused an epidemic of cholera or typhoid without first interacting in complex ways with a source of soil.[55]

As late as 1879 another colleague of Bastian's, Timothy Lewis, "was not convinced that bacteria can cause disease."[56] Lewis was working on cholera in India in particular along with D. D. Cunningham.[57] Cunningham was "an army surgeon attached to the Sanitary Commissioner with the Government of India" who favored the soil theory of Pettenkoffer. He did not embrace the soil theory without reservation, but he felt it accorded with the facts better than other theories. He stated this in 1871 in a report on the 1870–71 cholera season in the Madras district of India. Cunningham felt that in cases where microorganisms such as vibrios were numerous in cholera victims, they might have been produced by decomposition of the contents of the intestine.[58]

Not surprisingly, then, the response in the letters column of the *Times* to Tyndall's November 1874 letter on typhoid fever was largely from medical men and largely critical. Drs. Alfred Carpenter and C. R. Bree both emphasized the importance of the environmental factors that Tyndall was overlooking. Bree condescendingly pointed out that "every one knows, and has known for years past, except Dr. Tyndall—who, being non-medical, may be excused for not knowing," that Budd had a hobbyhorse in which he had "indulged most vehemently, but which has not in its totality been accepted by the profession."[59] An editorial soon afterwards in the *Practitioner* was equally indignant:

> Surely Professor Tyndall, when he put pen to paper to write his notable letter to the *Times* . . . must have done so under the belief that a capacity to understand medicine was in-born. In no other way can we comprehend a man of high scientific training in physics . . . committing himself unreservedly to a confessedly one-sided view of a recondite etiological question . . . It is not surprising, perhaps, that the learned professor of the Royal Institution . . . should have been profoundly impressed with the theory . . . of [William Budd]. But it is surprising that he should thereupon have assumed professional and public ignorance of the subject.[60]

Tyndall in his letter had written of a recent typhoid outbreak at a large institution at Over Darwen, claiming that Budd's theory could be easily put into practice there by using disinfectants as a panacea. The *Practitioner* sarcastically quoted: "he observes, 'Can it be doubted that, with sound medical advisers, backed by an intelligent population, an equally rapid destruction of the foe might be accomplished at Over Darwen?' Not only can it be doubted, but it may be affirmed that the vast accumulation of evidence on this subject . . . [is] almost wholly adverse to the efficacy of disinfectants under such circumstances."[61]

The writer said numerous experiments in public schools, jails, and other institutions had shown disinfection to be of little use in practical settings. The reviewer thought Tyndall had made Budd's simplified germ theory even more grossly oversimplified: "Unless the whole place could be sunk several feet beneath a sea of disinfectant, and whilst there its filth assiduously stirred up for some time so as to secure the sufficient action of the disinfectant on every part for some time, it is difficult to conceive how disinfection could be applied to it, except in name, on any large scale. Dr. Budd never intended that disinfection should be substituted for other fundamental means of limiting and preventing the spread of typhoid."[62]

It bears repeating that skepticism about the germ theory of disease and belief in a Liebigian model of zymotic disease production did not automatically imply support for or belief in spontaneous generation. William Carpenter did not subscribe to the germ theory, for instance, but was by no means a supporter of spontaneous generation. Yet in many cases medically trained observers found Tyndall's behavior so offensive that they were much more willing to provide Bastian with a forum for his views. This was also the case when Koch began to proclaim a very dogmatic link between specific bacterial species and specific diseases. Since in Britain many medical men were at least open-minded on, if not outright supportive of, pleomorphist views, their resistance to the early rigid formula of Koch provided at least some breathing room for Bastian's more extreme contradiction of Koch.[63]

Support in the Medical Community for Bastian

Although a large proportion of Bastian's audience may have been responding to rigid dogmatic formulations of germ theory, or to Tyndall's presumptive attitude toward doctors, nonetheless a significant number of medical men directly supported Bastian's theory of heterogenesis and even archebiosis. Henry Lawson of St. Mary's Hospital Medical School and the several popular science journals he edited have already been discussed, as has Clifford Allbutt. The *Lancet* of 9 July 1870 saw Bastian's first paper in *Nature* of the week before as a breath of fresh air—some real scientific experimental work, compared to Tyndall's lecture on "Dust and Disease," which this editorial represented as irresponsible blustering. The germ theory, said this writer, "has since been discussed, with more or less display of ignorance . . . by a large proportion of the general press. Amid or beneath all this turmoil, and quite independently of it, the work of real investigation has been going on; and Dr. Charlton Bastian has put forth, in *Nature*, what is perhaps the most important contribution yet made towards the solution of the great question that really underlies all controversies about germs. That question is, to determine whether there is any line of demarcation between animate and inanimate matter."[64]

The writer, probably editor James Wakley, went on to argue that the germ theory practically required spontaneous generation as its complement, since it could explain the rapid spread of epidemics in many cases but could say almost nothing about how they first got started: "The facts of life point almost irresistibly to the conclusion that epidemic diseases may originate *de novo* from certain combinations of conditions; and it is hardly possible to believe that the germs of all past or present pestilences were coeval in their beginnings with the peopling of the earth by man; and yet, if not, and if these pestilences do indeed depend upon living and organic self-multiplying poisons the doctrine of spontaneous generation would be at once established."[65] The *Lancet* writer also suggested that Bastian's papers in *Nature* were best viewed as trial balloons, intelligently used to gauge reaction to his theories, while he prepared his magnum opus, *The Beginnings of Life,* for press.[66]

The editorials and reviews in the *Lancet, British Medical Journal, Medical Press and Circular,* and *Practitioner* all were consistently favorable to Bastian's work in the early 1870s, and the *Lancet* and *British Medical Journal* continued to be supportive through early to mid-1877. In particular, all these journals gave very favorable reviews of *The Beginnings of Life* shortly after Macmillan published it in the summer of 1872. The *Practitioner* declared "we may decidedly say that to unbiased outsiders it will seem that he has put M. Pasteur and the rest of his opponents in a very ugly and difficult dilemma. Certainly, a large portion of his criticisms of Pasteur are excessively damaging, nor do we at present see how they can be less than fatal, to the reputation of the latter as an experimenter of high rank. It is, of course, quite another question whether Bastian, in his turn, has exercised due care; we can only say that we perceive no flaw."[67] The writer went on to laud Bastian for his extraordinary ability to render the complex theories of life and energy accessible, even to the average reader. Not since Maudsley's 1863 review on vitality in the *British and Foreign Medico-Chirurgical Review,* he said, had this author seen as up-to-date, "brilliant," and well written a summary of these issues as in Part I of Bastian's book.[68] Furthermore, this writer felt Bastian had been successful in his attempt to generate new terminology that escaped the outdated connotations of the old term "spontaneous generation":

> Dr. Bastian has done very necessary work in clearing up a natural, but very unfortunate, confusion in the ideas of those who talk popularly

> about "spontaneous generation." He shows very well, first, that the word "spontaneous" is useless and unscientific; and secondly, that under that popular phrase two ideas are confounded, only one of which has (until now) been seriously advocated since the birth of scientific biology . . . The audacity of Bastian carries him beyond the doctrine of heterogenesis: he believes in the . . . occurrence of what he calls . . . "Archebiosis" . . . Once the first shock of the idea is got over, however, . . . the reader will find far less to astonish him in the . . . evidence produced by Bastian . . . to an extent that no previous Darwinian philosopher has attempted to prove.[69]

The reviewer in the *Practitioner* wholly sympathized with Bastian's position. He was carrying the new physicalist science to its logical conclusions, opposing vitalism. And the reviewer said Bastian was right that the opposition of "a few prominent physiologists" was due to their fear that this line of reasoning "must land them in the regions of atheism and materialism . . . The brunt of conflict which Dr. Bastian will have to sustain, however, is not with the dwindling sect of vitalistic biologists, but with the far larger and more influential section of scientific men [i.e., Tyndall and Huxley] who candidly acknowledge the substantial identity of the forces that originate life with those of the physical world, and yet cannot bring themselves to think it even possible that living things should originate from not living matter."[70]

The *Lancet* review was strongly positive and long enough to split into two installments, on 12 and 19 October 1872.[71] The *Medical Press and Circular* review was equally glowing, and the writer was so convinced by Bastian's argument of the inseparable link between evolution and his work that he used the terms "heterogeny" and "evolution" quite interchangeably. The reviewer went on to opine that "few but unqualified observers or shallow pathologists could continue long to entertain the idea that the spreading diseases are of parasitic origin. This remarkable minority, nevertheless, exists."[72] Furthermore, he suggests that supporters of the germ theory seem all to be committed to the use of carbolic acid, with a faint implication that this may be a conflict of interest: "in this country the principal supporters of the germ pathology are an eminent surgeon who has played a prominent part in popularising antiseptic surgery by means of carbolic acid; a London physician, who claims to have made a valuable discovery in recognising the therapeutic properties of the sulpho-carbolates; and a provincial Professor, who is best known as a manufacturer of carbolic acid."[73] The surgeon was Lister, the London doctor was Arthur E. Sansom, and the manufacturer was F. Crace Calvert.

The *British and Foreign Medico-Chirurgical Review* was more measured, but still declared in April 1872 that "Dr. Bastian is certainly an adroit controversialist, as well as an innovator. We remember, last year, a violent polemical discussion which was carried on between him and Professor Huxley. In the inception of the dispute the facts were all on the side of the Jermyn Street professor; but, owing to the skill of his antagonist, . . . Dr. Bastian carried his lance fairly broken out of the arena."[74]

The writer cited experiments by Burdon Sanderson in which no life developed in infusions, as long as the vessels used were thoroughly sterilized first, and went on to wrestle with the ambiguity of the controversy: "when two such skillful observers arrive at such diametrically opposed results, we, at least, must be pardoned for deferring our judgment . . . It is true that, because [Bastian's] testimony is at variance with that of other observers, it need not necessarily be wrong. There have been many instances, even in anatomy and physiology, of the opinion of one solitary observer being opposed to the unanimous voice of his contemporaries, the one man having been afterwards triumphantly proved to have been correct."[75] And, while uncomfortable about archebiosis (abiogenesis) and about Burdon Sanderson's inconclusive experimental results, the writer muses

> Were it not that we are investigating a strictly scientific subject, on which the mere facts have to be examined and taken at their value, we would be inclined to think that the probabilities in favour of the origin of living beings, as Professor Owen has pointed out, by a sort of heterogenesis, has much to be said in its favour. We regard the alleged refutation of the probability of heterogenesis, made by Professor Huxley at the Liverpool meeting, as entirely unsatisfactory, based, as it was, upon not a single cited original experiment. The advocates of spontaneous generation have a right to demand a demonstration of the impossibility of their statements, instead of a mere allegation of their improbability. Dr. Bastian has certainly carried out a long series of experiments . . . and the character both of his experiments and of his views is such as to call for a serious re-examination and discussion.[76]

Charles Murchison issued a second edition of his widely respected text *On the Continued Fevers of Great Britain* at this time. He, too, gave cautious support to Bastian in the opening pages of his text.[77] As mentioned previously, just a few months later, in August 1873, Murchison's reputation grew enormously when he almost single-handedly tracked down the cause of an alarming typhoid outbreak in the wealthy neighborhood of Marylebone, where many noted London physicians had their practices. The stature of Murchison's skeptical view toward a simplistic germ theory grew accordingly. The *Daily Telegraph* noted:

> A great number of most respectable families—living in houses not only good, but of the first class—have seen typhoid fever stalk into their midst . . . As many as one hundred and sixty-five persons . . . have been affected with this "disease of dirt," in such localities as Wimpole Street, Harley Street, Cavendish, . . . Here were a dozen West-end doctors—some of them of high reputation as authorities upon fever . . . The fact is noteworthy that forty out of forty-three households, in some of the best parts of London too, should all have typhoid and all be drinking the same milk . . . and it would hardly remain to ask any other question than "how did the milk come to be poisoned with the typhoid zyme?" . . . No microscopist has seen . . . the subtle spore or monad which does this deadly mischief . . . [T]his is not yet demonstrated . . . On the other hand, milk, by reason of its oleaginous and albuminoid components, is a very possible vehicle for contamination.[78]

The writer clearly considered the germ theory unproven at best and the zymotic theory more likely as a scientific replacement for the outdated notion of "dirt." Thus, during the years from 1872 until at least 1876, Bastian's theories flourished in a generally sympathetic environment in the mainstream medical community. Furthermore, the *Telegraph* considered the entire area of research so important that it repeated its suggestion "that the whole question of zymotic disease should be taken up by Government at the national expense. The subject is ripe for a searching and persevering prosecution by a mixed commission of scientific, medical, and physiological men . . . Human lives by the ten thousand hang upon that query."[79]

Huxley had effectively begun to restrict Bastian's use of the Royal Society as a platform for his views by mid-1873, and Lankester had consistently used his editorship of *Quarterly Journal of Microscopical Science* as a weapon against Bastian. Nonetheless, during this time the Pathological Society and other medical groups remained sympathetic venues for Bastian's opposition to any "hard" formulation that germ = disease, and all of those groups considered that his research was an important part of the overall campaign of science to attack the problem of communicable disease.[80]

The Pathological Society Debate of April 1875

In the various London medical societies, the new evidence about the nature of bacteria was being disputed actively during these years. In January 1874, a lengthy dispute broke out at the Clinical Society of London following a report on the possible involvement of bacteria in pyæmia (septicemia), in which both Bastian and Burdon Sanderson took part, both being founding members.[81] The debate continued through the next four meetings of the Society, sometimes growing heated. Thus, when the Pathological Society of London announced a year later that a general debate was to be held on the germ theory of disease, there was considerable interest. The membership of the two Societies overlapped somewhat, and Bastian was to open the discussion by presenting a detailed critique of the germ theory for an hour, i.e., most of the opening session of 6 April 1875.[82] Burdon Sanderson was then asked to give the first response, leading off support for the germ theory, which he had increasingly defended publicly since 1870.[83] Then, at the next meeting of the Society, on 20 April, the debate was opened to the floor, with contributions from Dr. T. J. Maclagan of Dundee; another Scot, Dr. John Dugall; Dr. E. Crisp of Chelsea; Mr. Jonathan Hutchinson, surgeon to the London Hospital; and Mr. Knowsley Thornton.[84] Dr. Charles Murchison agreed to open discussion at the third session on 4 May 1875, and he was followed by Mr. W. Wagstaffe, Dr. J. F. Goodhart, Dr. Payne, and Mr. Jabez Hogg.[85] Bastian then gave response, rebuttal, and summed up the discussions.[86]

In his initial paper Bastian began by stating that the analogy between fermentation and zymotic disease historically had been considered very strong, so that now that Pasteur was widely believed to have shown fermentation to be a living process, many doctors were leaning toward believing in the germ theory of disease. So strong was the analogy, he argued, that as many were convinced in 1875 that germs were the true contagia as had been convinced of "the opposite notion, founded upon Liebig's physico-chemical doctrine of fermentation, some twenty years ago."[87] This was premature on two levels, he insisted. First, he asked, are we justified in relying so heavily on the analogy between zymosis and fermentation in simple mixtures such as sugar solutions? Second, he suggested that Pasteur's claims that ferments can *only* be living organisms were not as fully conclusive as they had been regarded to be. For instance, mere particles or fragments of organic matter had recently been shown to initiate some fermentations. But while that point was still under debate, he concluded that meanwhile "we are free to look into the question of the relation of the lower organisms to disease on its own merits—apart, that is, from the overweening influence which might be exercised by any generally accepted theory of fermentation."[88]

Next, Bastian pointed out that the germ theory was not a single unified approach; that there were at least two significantly different versions of it lumped under that name, namely those of Burdon Sanderson (similar to Pasteur) and of Beale. Lionel Beale's theory held that the particles of contagion were indeed particles of living matter capable of reproducing themselves, which he called bioplasts. He claimed, however, that these bioplasts were the products of degeneration in the tissues of a diseased organism. These then became transmissible and capable of setting up the disease in a newly infected organism.[89] Beale, like Burdon Sanderson, had begun his researches on the subject of contagion as part of the Government Commission on the Cattle Plague of 1865–66. However, by 1869 or 1870, despite Burdon Sanderson's attempt to minimize their differences, the two men had fallen out over their interpretations of the experiments each conducted.[90] The multiplicity of germ theories, and the internal disagreements over important points, were a weak spot in the opposition to zymosis via heterogenesis/archebiosis, and Bastian utilized this weakness to persuade medical men that the "germ theory" was not the solid, consistent doctrine that many of them supposed.

Furthermore, he went on, both Burdon Sanderson's and Beale's germ theories were unproven. True, Bastian said, they had shown that an increase does occur in numbers of bacteria and "particles of bioplasm" in diseased animals, but neither of them had shown that it was one of those particles that was the contagium that set the process in motion. Thus, it was equally possible that these microscopic living things were *products* of the disease process. There were, Bastian argued, at least four distinct reasons why the bacteria could not be the causes of illness: 1) They could be introduced into the blood vessels of experimental animals by the thousands, yet produce no ill effects in a large proportion of those inoculated. 2) Bacteria exist in myriads in many different tissues of the healthy human body (even if not actually found within the blood vessels in health). 3) The usual reply to the first two problems was based on denying the widely accepted fact of pleomorphism:

> It is no answer to such difficulties to say that there are distinct species amongst these lower organisms, some of which are harmless though others are poisonous . . . Against this interpretation may be brought the experiments of several investigators, showing that bacteria are mere creatures of circumstance, and modifiable to an extraordinary degree. The last position is even admitted by Professors Sanderson and Lister . . . Lister's own work has compelled him to make an admission, which in the face of . . . the wide distribution of bacteria within the body, seems fatal to any belief in the germ theory of disease.[91]

He then cited Lister's affirmation of pleomorphism from the latter's paper of the October 1873 *Quarterly Journal of Microscopical Science.* Bastian's ability to find data in his opponents' papers that allowed of an interpretation favorable to his views was a constant source of irritation to them. As we shall see, this was given a very unflattering interpretation by Tyndall as it showed itself over time to be a highly effective tactic.

The fourth point Bastian made was that Lewis and Cunningham, Burdon Sanderson, and others had established two experimental facts completely at odds with a germ theory. They had shown that some contagious fluids decreased in virulence as the numbers of bacteria in them actually increased. Further, some contagious menstrua had been shown to lose scarcely any of their contagious properties after boiling at 100°C in the moist state for a few minutes.[92]

Because all the evidence seemed to be against germs as the cause of disease, Bastian argued, their relation to illness should be considered to be other than causative: "I maintain, in short, that even the very existence of organisms in the fluids and tissues of diseased persons is for the most part referable to the fact that certain changes have previously taken place (by deviations from healthy nutrition) in the constitution and vitality of such fluids and tissues, and that bacteria and allied organisms have appeared therein as pathological products—either by heterogenesis, or by . . . archebiosis."[93]

Because the germ theory at this time did not present a united front, because pleomorphism was a widely accepted view of bacteria, and because germ theorists still lacked any sufficient concept of immune response to explain the data Bastian cited, he was believed by many medical men to have offered at least as good an account of the facts, including a role for the patient's constitution, with his version of zymotic theory. He had convinced many that the facts were simply too complex and refractory to support a simplistic version of the germ theory. None of his opponents in the debate could deny that Bastian had as masterful a command of the most recent experimental science bearing on the question as anyone involved. Indeed, he was lauded for his encyclopedic knowledge and for his ability to synthesize the myriad facts to build a strong case.

The Physiological Society

The anti-vivisection campaign in Britain was growing rapidly in the mid-1870s, and all the up-and-coming medical scientists were concerned lest its influence should lead to laws against animal experimentation. Initiatives against such legislation were being led by Burdon Sanderson and Michael Foster, with Huxley, Hooker, and even Darwin occasionally joining in to show support and solidarity among the "Young Guard" in the evolution-based sciences.[94] By early 1876 a core group of those whose research was most directly affected decided to meet and form the Physiological Society of London, to further the cause of physiology and especially to coordinate the defense against the anti-vivisection threat. My argument is that the clash over spontaneous generation in this context, and the polarization as the medical journals supported Bastian and opposed Tyndall, contributed to a consolidation of Victorian laboratory science as a community more distinct and separate from clinical medicine than would otherwise have been likely.

By this moment in its development, laboratory science had reached a significant milestone toward the professionalization and respectability that was the ultimate goal of men like Huxley. And all the laboratory men of the Young Guard were there. All, that is, except one. This moment in the development of Victorian laboratory science, epitomized by the "snapshot" of the signatures of those present at the opening meeting, shows the conspicuous absence of one of the most prominent public voices who shared in the agenda of protecting and advancing medical and laboratory science. Henry Charlton Bastian had been a founding member of the Clinical Society ten years before and was a respected name in the Pathological Society and the Microscopical Society. He had even been elected to the most elite of these groups, the Royal Society of London. But that was eight years previously, before he had so openly defied the authority of Huxley and refused to "eat his leek." Now, although most of his University College medical colleagues were invited by Burdon Sanderson to join the new society on 31 March 1876, including John Marshall, Edward Schäfer (later Sharpey-Schafer), Sidney Ringer, and George Thane, Bastian was not invited.[95] Soon, Ray Lankester in Zoology was invited to join, as were Ernest Hart, editor of the *British Medical Journal,* Darwin's young protégé George Romanes, and even young Victor Horsley and William Halliburton, just recently medical students in Bastian's classes. But not Bastian. Neither the Society's official minute book nor Schäfer's history of its first fifty years mentions his name, not even to justify or explain why he was not invited.[96] Given Bastian's prominence as an advocate for medical laboratory science, there could be no more eloquent testimony to his fall from grace among his colleagues than such complete social exclusion from a group to which one would fully expect him to belong. Clearly Bastian had stepped beyond the bounds allowed for gentlemanly disagreement, most of all by airing that disagreement so publicly. Burdon Sanderson, by comparison, was considered incompetent and worse by many, but not in a way that called for ostracism or exclusion, perhaps because he did not air his disagreements publicly, for the most part.[97]

It might seem that Bastian had spent all of his scientific credibility by this point, but this was not true. He was shortly to become the Dean of the University College Faculty of Medicine.[98] Edward Youmans, the pro-Darwin American publisher, ran a biographical feature on Bastian and his work in the "scientist of the month" column of his *Popular Science Monthly* in November 1875. Even the Royal Society still called on him as a reviewer for the *Philosophical Transactions,* at least on papers in neurology.[99] More important, Bastian was recruited at this time to write one of the texts for the International Scientific Series. The publishers of this series were eager to forward the cause of the evolutionary Young Guard, especially by creating basic texts in different areas of the sciences for a mass audience, each written by one of the top authorities in its field.[100] The English committee of scientists advising the publisher included Tyndall, Huxley, and Spencer. Bastian's stock was still high enough that he was offered a contract for *The Brain as an Organ of the Mind* late in

1875. He was given a £100 advance, and the publisher contracted for a first run of 2,500 copies, twice the usual number for the series, on expectations that the book would sell very well. Spencer, whose close friendship and respect for Bastian as a scientist remained intact throughout his lifetime, may have championed Bastian's cause on the scientific committee, since Huxley and Tyndall would likely have opposed it.[101] Bastian contacted Richard Owen, who was by this time heading up the creation of a Natural History Museum in South Kensington, to ask Owen for use of some of his published illustrations in the new book.[102]

By December 1876 the new Physiological Society had enough cash in its coffers to add small research grants to its functions. It invited grant proposals from members, and one of the first to come in was from Lankester. He proposed to work on two major topics, invertebrate embryology and "the natural history of the organisms concerned in putrefaction."[103] Furthermore, though the invitation had specified *small* grants only, Lankester asked for £200 per year for five years. The committee on grants decided that he was asking for a single grant for research on two very different topics, and they recommended against supporting both. They went on to say, however, "if Mr. Lankester should think fit to apply to the Grant Committee for a sum of £200 for five years to enable him to continue his researches on the development of the invertebrata, your Committee believe that the importance and the difficulty of the proposed enquiry, the proved and exceptional qualities of the investigator and the character of the work which he has already published on the same subject, would justify the Society in cordially supporting the application."[104]

The Society was willing to support Lankester's research, despite his personality conflicts with both Burdon Sanderson and Schäfer, which at the time were almost as bad as those two scientists' relationship with Bastian.[105] The Society seemed disinclined to support even one of its own, however, on research that was clearly aimed against Bastian, perhaps wishing to stay clear of any taint of involvement in the spontaneous generation controversy or in the bad blood that it had created among the evolutionary community.

In the years that had elapsed since 1870, Huxley's course in elementary biology had begun to bear fruit. The course acted as a unifying common training ground for the younger generation rising into the London biological societies. One historian has observed that, between 1870 and

1878, "the flower of a whole academic generation passed through this South Kensington filter, as imbibers of the course, or more important, as its expounders, and went on from there to instill in the universities, in the training colleges, and in the schools the new experimental approach and the novel theoretical concerns that gave to the then incoming tide of learning an appearance sharply different from all that had preceded it."[106]

Not surprisingly, in Huxley's elementary biology course, and in all the spin-offs taught by Lankester, Dyer, Foster, Newell Martin, and others, there was no place for Bastian or his experiments. In the manual used as the textbook for these courses, abiogenesis was authoritatively confined to the earth's distant past, and bacterial generation was said to be by fission or by spores only.[107] The influence of this course, upon which the training of all future generations of British biologists was modeled, can not be underrated in spreading the Huxley/Tyndall version of evolution. Bastian's story had no comparable disseminating mechanism to new generations of biologists. He could reach only the audience of his medical students.[108] A few biologists continued to support Bastian on the Continent, as did the Russian revolutionary Alexander Herzen.[109] But this could in no way keep up with the whole new generation being trained in the Huxley tradition. The sudden death of Charles Murchison in April 1879 removed one of the very last figures that was skeptical of the germ theory and yet had still been considered "respectable" as a researcher. This tipped the balance further against any possible rapprochement between the laboratory research community and the medical germ theory doubters.

As the spontaneous generation controversy progressed in Britain, the rhetoric of Tyndall offended medical scientists and doctors from his first public lecture in 1870. Their response, including Bastian's, was sharp in tone, which Tyndall resented, and the grievances escalated steadily. The result was an increasingly sharp polarization of the debate along medical/nonmedical lines, with difficulty in finding a middle ground on spontaneous generation for any who were committed to both clinical and laboratory work. Doctors committed to evolution found that the polarization between the X Club and Bastian left a similar no-man's land in between. However, a significant number of them seemed convinced, by Bastian's experiments as well as his logic, that the germ theory was still a poorer explanation of disease than the zymotic theory. Many of them also found his version of evolution more internally consistent, and accused the Huxley camp of trying to avoid spontaneous generation only because they wished not to deal with its materialist consequences. The polarization along medical/nonmedical lines had important consequences for the character of the newly crystallizing professional discipline of physiology, causing that discipline to exhibit a sharper separation from clinical medicine at the time of the founding of the Physiological Society than might otherwise be expected. Teaching the death of spontaneous generation in fact became one of the foundations for training in physiology, through Huxley's elementary biology course.

7

Purity and Contamination: Tyndall's Campaign as the Final Blow

In 1974, Friday, MacLeod, and Shepherd complained that John Tyndall's role in settling the spontaneous generation controversy had been given too little attention by historians.[1] Works by Friday, Farley, and Adam that have appeared since redress this imbalance, giving Tyndall some of the credit they felt he had been denied.[2] There is a larger sense, however, in which the accounts have still not been put to rights. And that requires a telling that portrays the intense level of activity Tyndall engaged in *beyond* the laboratory and the normal public venues of science. It would be no exaggeration to say that the momentum of Bastian's successes was opposed, and finally brought to a halt, by no other single force as active or as concentrated as Tyndall's personal campaign against him. And through Tyndall's personal influence, the work of Koch, Cohn, and Pasteur, which had been previously fairly marginal to the British debate, was brought to bear directly against Bastian. When this experimental work alone failed to turn the tide, and when, to Tyndall's horror, Pasteur even complimented Bastian on his experimental prowess, the British physicist labored with redoubled zeal for almost a year to convince Pasteur that Bastian was a shameless opportunist. Further, Tyndall extracted pronouncements from Pasteur that opposed Bastian's theories, making sure that these were worded in ways calculated to deny Bastian any tactical advantage. During this time Tyndall also attacked Bastian, both personally and with regard to his scientific talent, to any prominent British scientist who would listen, especially to Darwin and the X Club. By the mid-1870s, few evolutionists failed to get the message: evolution and Bastian-style spontaneous generation were not only to be seen as not mutually implied, but indeed as incompatible. Network building was the key to Tyndall's success—a story that has yet to be told. In this sense, Tyndall's own construction of the debates over spontaneous generation as a textbook *experimentum crucis* has yet to be seriously challenged.[3]

Much more was at stake for Tyndall than merely a question of experiment. From Tyndall's point of view, for instance, the overall state of the medical journals was to be judged as deplorable because they supported Bastian's experiments and conclusions. Furthermore, the current level of education of the British public with respect to science was to be gauged, according to Tyndall, by whether or not they could see, as he did, that Bastian was a complete charlatan. The motivation for Tyndall's stance on spontaneous generation has long been problematic for historians. As Farley put it, Tyndall "presents to the historian a confused panorama of beliefs: evolutionist, materialist, germ theorist, and opponent of heterogenesis."[4]

Tyndall's A Priori Commitments

So why would Tyndall be automatically opposed to spontaneous generation, a position even more radical than that taken by Lister? One early reason may have been Owen and Pouchet's defense of it, combined with their general antagonism to Darwin's theory and, in Owen's case, with opposition to a Daltonian theory of atoms. Tyndall was firmly and publicly committed to atoms as real material objects in Dalton's sense, at a time when major debates on this point were occurring at the Chemical Society of London in 1867 and 1869. Owen, however, attempted to link his spontaneous generation theory, when finally published in detail in 1868, with the opponents of discrete atoms in an Oken-style view of matter as purely a secondary manifestation of force. Owen cited the authority of the recently deceased Michael Faraday, an entirely plausible reading of some of Faraday's later and more speculative writings on the subject, in which Faraday referred to atoms as "centres of force."[5] In his biography of Faraday published in late 1869, Tyndall's close friend Dr. Henry Bence Jones noted these speculations of Faraday's. Perhaps in reply to Owen and other anti-Daltonians, however, Bence Jones quoted from Tyndall's *Faraday as a Discoverer,* which tried to downplay the importance of such ideas, citing them largely as proof that Faraday was not above speculation.[6] Tyndall's own belief in atoms was fervent and realistic, so much so that Huxley kidded him that atoms had become for Tyndall the same kind of emotional, faith-based ontological foundation that God provided for others.[7] In his own biography of Faraday, Tyndall relayed Faraday's statement that the atom was a concept that tempted scientists to go beyond the actual facts of combining proportions, gas volumes, etcetera, and to invest it with ontological reality it did not deserve. Tyndall replied, "Facts cannot satisfy the mind; and, the law of definite combining proportions being once established, the question 'Why should combination take place according to that law?' is inevitable . . . The objection of Faraday to Dalton might be urged with the same substantial force against Newton: it might be stated with regard to the planetary motions that the laws of Kepler revealed the *facts;* that the introduction of the principle of gravitation was an addition to the facts."[8]

Tyndall's reply to Faraday's skepticism about atoms could be copied word for word to represent his reply to the doctors' skepticism about germs. For Tyndall, Lister's results with antiseptic surgery functioned like Kepler's laws, revealing the facts of the matter; Pasteur's germs functioned like Newton's gravitation, supplying the more basic principle of nature.

Bastian, interestingly, did *not* try to link spontaneous generation with Faraday's or Owen's type of matter theory. He fully embraced Darwin's evolutionary theory (having no contact with Owen at all, prior to 1876), and his view of the "correlation of forces" and atoms as real physical objects was modeled on the writings of Tyndall during the late 1860s.[9] It was the logical consequence of precisely that reasoning by Tyndall, by Huxley in his "Physical Basis of Life," and by Bence Jones in his *Croonian Lectures*[10] that Bastian claimed should lead one to see archebiosis as possible. Indeed, Tyndall himself often emphasized the extraordinary self-organizing power in inorganic matter, both in his writing and in his flamboyant public lectures. He often cited, for example, the fact that water and salt molecules could form exquisite crystal structures that rivaled the forms of the animal and plant kingdoms.[11] Before 1870, many believed crystal structures implied that the boundary between life and nonlife could be crossed. If the powers inherent in matter did not extend to forming primitive living protoplasm, Bastian argued, the only alternative was to posit a vital force to explain why living things should not be able to form by a process analogous to crystallization. Yet vitalism had been specifically opposed by Huxley in his lecture, "The Physical Basis of Life." And Bastian agreed fully:

> if the "vitalist" wishes to establish the existence of a more fundamental difference between crystals and organisms than we are prepared to grant, seeing that the scientific evidence seems to be against him, it remains for him at least to endeavour to show good grounds for the establishment of such difference. It should be remembered, then, that in the present state of science all theoretical considerations seem favourable to the views of the evolutionists, and the only thing which can be opposed to them is the *assumption* that those processes of reproduction which take place amongst all . . . living things are the *only* processes by which such living things can arise.[12]

Bastian's conclusion that spontaneous generation was a logical implication of evolutionism was widely believed, as I have already shown, by both supporters and opponents of evolution. And Bastian's star was rising, while Owen's was distinctly on the wane.[13] Thus, Bastian's claim of spontaneous generation was dangerous to Darwinism in a way that Owen's was not.

Tyndall's negative response to Bastian was also related to another long-standing commitment. Almost ten years earlier the physicist had gone on record (in a letter to Darwin) on the importance of public science education, saying, "I think I could promise to perform my share of the business. I think it may become a very important agency and exercise a salutary influence in this quack ridden country."[14] Tyndall shared the frustration of Huxley and the X Club, during the intervening decade, when numerous attempts to establish a regular science journal that spoke for the "young guard" all failed, leading many to conclude that "science is not sufficiently advanced in England, notwithstanding the labours of our scientific men, to ensure for such a paper . . . at present popular appreciation and patronage."[15] In Tyndall's rhetoric, after 1870 the public's sympathy toward claims about spontaneous generation was said to represent an outstanding example of how British science education was failing and lagging behind.

Yet Tyndall did not start out believing that either Bennett or Bastian were quacks, or anything other than honest, sincere, able men of science. Indeed, through early 1870, while Spencer, Huxley, and Tyndall all made public statements opposing spontaneous generation, they all (especially Spencer and Huxley) still seemed to want to keep an open mind on the subject, since the implications of *Bathybius* and of the correlation of forces were not lost on them.[16] Although criticizing the theoretical conclusions of Bennett, for instance, of his experimental *results* Tyndall in January 1870 said, "their harmony with the conditions now revealed is a proof of the honesty and accuracy with which they were executed."[17] It is crucial that during this early period Tyndall also saw Bastian as very talented, sincere, and respectable. In a letter after his January 1870 lecture, Tyndall discussed an apparent plan by the two to meet and compare experiments, saying he needed to postpone for a few days. He added, "But you probably think as I think, that it will be better for each of us to pursue his independent way. I do, I confess, entertain the very earnest wish that such evidential value as my experiments may justly claim will be made clear to your mind. It would be no pleasure to me to see so able a man ranged against me on so grave a question."[18]

It was only after Bastian confronted Tyndall in an exchange of letters in the *Times* in April 1870, sharply challenging a physicist's authority to make claims to understanding disease processes better than doctors, that Tyndall turned against Bastian. Tyndall seems to have been shocked by what he considered Bastian's outrageous rudeness, humiliating him in such a public forum. And this greatly changed his perception of Bastian as a gentleman and ally. As we have seen earlier, Tyndall's own experience as an experimenter and a medical patient contributed to his commitment to the germ theory. Now the confrontation with Bastian further stiffened Tyndall's already considerable resolve on this point.

To Huxley he wrote that John Burdon Sanderson was one of the few forward-thinking medical researchers worthy of praise, but of Bastian, Tyndall concluded, "his impudence is certainly astounding."[19] And by late 1873 he was spreading his feelings beyond X Club intimates, as in a letter to Frederick Barnard, president of Columbia College in New York, whom Tyndall had met on a recent lecture tour of America: "I am not surprised to find you attaching so much importance to Bastian's work. Still there is not a man of my acquaintance of any scientific weight, and I number among my acquaintances many who know Bastian's calibre and method of work, who attach any importance to his results. All his more startling ones are to be ascribed to the fact that a man undisciplined in experiment has taken up a subject which requires for its treatment the most consummate experimental tact."[20]

It is interesting to contrast this with Tyndall's assessment of Bastian before the younger man had publicly confronted him and before Huxley had changed his mind about Bastian's technical skill. Of their earlier esteem for Bastian as a scientist no trace is to be seen in their later assessments, including those public accounts on which subsequent histories in science texts have been based. In Tyndall's statements, as with other foes with whom he clashed publicly, his opponent became increasingly demonized as time passed.[21] So did all who supported the opponent. Thus, by 1875, Tyndall concluded that the medical profession itself was its own worst enemy and obstacle to progress.

Embarking on the Quest and Recruiting Support

Since Bastian had by that time become so visible as the major proponent of heterogenesis and archebiosis in the public eye, it was Bastian in particular upon whom Tyndall focused all his energy when he decided in mid-1875 to embark on new experimental work on the subject. As he expressed his aim a few months later to Huxley: "I must live the life of an ascetic, to clear away this 'Bastian fog.' It is amazing what audacity can do in England; & his audacity has powerfully influenced numbers of intelligent people. Without entering into controversy with him I hope to set him in his true light. I have been re-reading his criticisms of your Liverpool address—they almost take one's breath away . . . Something, I think, may be done to dispel this illusion."[22]

And Tyndall continued the same moralizing description of his view of the contest in a letter turning down all social engagements:

> At present I fear anything that interferes with my work, which demands the most calm and concentrated attention: and I am quite determined for a good while to come to hold on loyally to it and to set my face like a flint against all social temptations . . . my work must go on without interruption. Pitfalls and enemies are before me and around me, and I am resolved not to be tripped up by the one, nor overcome by the other. But failure can only be avoided by making every part of my investigation unassailable. This, God helping me, as old Luther would say, I intend to do, and this can only be done by solitary thought and severe experiment.[23]

The escalation of moralistic language in describing his own role of "knight taking confession before embarking on the holy quest" corresponds to a similar escalation in the level of villainy attributed to Bastian. A cartoon drawn two years later, found among Tyndall's papers (see figure) details some of the stages on his quest (with considerable humor), and reveals imagery of similar moral content: the germs are ultimately Victorian demons, who frolic and revel in the vials of Dr. Bastian, but are chagrined at Tyndall's own unassailable vials. If the artist was not Tyndall himself, it must have been a sympathetic friend who knew the controversy in considerable detail.

In his quest to vanquish Bastian, Tyndall sought to enlist the most powerful authorities he could, starting with the biggest of them all: Pas-

Cartoon (possibly drawn by Huxley) illustrating the experimental clash between Bastian and Tyndall, 1875–1877. Found among Tyndall's papers and reprinted courtesy of the Royal Institution. The menu of the "Pasteur Inn" features "for Dinner, offered by Messrs. Tyndall and Pasteur for the Academy of Sciences," dishes such as "Soups: Clear Hay, Bacterious Hay; Fish: macerated salmon with cucumber sauce," all substances in the experimental infusions. Hay infusion and cucumber infusion had produced particularly contested results. The menu also offered "Petits Rotifères au Diable."

teur. Tyndall had first corresponded with Pasteur in May 1870, just after his initial clash with Bastian in the *Times*.[24] Upon Pasteur's suggestion, Tyndall immediately set about studying the French savant's treatise, *Diseases of Silkworms,* and he wrote a summary and very glowing review of the book for *Nature*.[25] In November 1874, Tyndall was instrumental in getting the Royal Society to award Pasteur the prestigious Copley Medal, for which Pasteur expressed his personal gratitude.[26] And following the April 1875 debate in the Pathological Society, in which Bastian's skepticism toward germ theory seemed basically supported by much of the medical community, Tyndall began to bring Pasteur up to date on Bastian and his great influence in Britain. Simultaneously, he set about trying to persuade the French chemist to lend his name to Tyndall's campaign against Bastian:

> I herewith send you the British Medical Journal, containing an article by Dr. Bastian. To this article I am now writing a reply. He has published the same article in the Lancet. It will, I think, be better for you to wait until my reply appears. It shall be sent to you on Saturday next. Meanwhile, I will so revise my letter as to render it fit for publication. I hope to be able to express with greater clearness my opinion of the relative merits of you and your opponents. You will see that Dr. Bastian takes the liberty of citing you as a supporter of his results. I wish you would send me two lines stating whether you consider him justified in thus citing you. It was high time to put a stop to Dr. Bastian. He was doing incredible mischief among the medical men of England and America.[27]

Pasteur agreed to come on board in the limited sense that he sent a letter that Tyndall could use for publication, but he did *not* believe Bastian was trying to be deceptive or to misrepresent his results.[28]

Tyndall also set about recruiting other allies and witnesses. He visited Charles Darwin as he began his experiments, and when the *Lancet* and *British Medical Journal* took a less negative tone toward those experiments, he wrote to Darwin: "To my regret, I shall be forced to go into the whole history of Bastian's work, dealing with his logic as well as with his experiments. I was disposed to deal with him in the tenderest manner; but his recent exhibition in the Times shows me that a far different treatment will be needed. The change in the Medical Journals is radical—they see that the end of the nonsense which they have so long countenanced is nigh."[29]

To which Darwin replied: "I am particularly glad that you are not going to give up at once the Spont. generation question. I do not care much about what Dr. Bastian says, but I feel *very strongly* that the whole subject is not made clear until some light is thrown on the question how men like Burdon Sanderson and Wyman of Boston and Dr. Child often succeeded in getting bacteria in infusions which they had boiled for a long time. Do for heavens sake complete as far as possible your beautiful work. I have as yet read only the first part in *Nature*."[30]

In his paper to the Royal Society on these experiments, Tyndall reported sending some of his tubes to be exposed and observed by many independent observers at different locations. These included Darwin and his son Frank at Down, Sir John Lubbock's study, Mr. Siemens at Sherwood, Mr. Russell at Pembroke Lodge, the daughter of Lord Claud Hamilton (soon to be announced as Tyndall's fiancée) at Heathfield Park, Sussex, Mr. Hirst at Greenwich Hospital, Dr. Hooker at Kew Gardens (several locations, including the orchid house), Mr. Price at the Crystal Palace at Sydenham, and numerous tubes examined closely by Professor Huxley at South Kensington.[31] Though the participation of most of these worthies did not add to the science in any significant way, such an illustrious list could not but impress the scientific, the aristocratic, and middle-brow audiences alike, with the sheer extent and variety of Tyndall's network. He was invoking numerous impressive places known to the public as images of respectable science (and Government support for it), in addition to the scientists themselves. By their sheer number and variety, these allies would surely outweigh whatever power bases seemed to back Bastian, such as his University College Hospital and Medical School positions, and the medical journals and societies. Even the very title of Tyndall's paper ("On the Optical Deportment of the Atmosphere . . .") seems to invoke respectable science taming the unruly behavior of atmospheric dust, grading it on its "deportment," as it were. Tyndall the master showman of science would subdue it to do the bidding of Science, just as he had taken control of the blue of the sky and used it in his public demonstrations. The title also reflects Tyndall's determination to show up the unruly deportment of Bastian and the medical journals, by portraying himself and his experiments as the epitome of dignified, disciplined, morally superior and intellectually lofty Science in the service of Truth, with the "pure, limpid, uncontaminated infusion" as its emblem.

The Rev. W. H. Dallinger was also recruited to Tyndall's cause at the last minute before going to press. His study of the life cycles of protozoa ("monads") was discussed in chapter 5. Dallinger had contacted Charles Darwin in January 1876. He had then been put in contact with Tyndall by Darwin, who forwarded Dallinger's letter to Tyndall,[32] and also by Henry Lawson.[33] As described in chapter 4, Lawson had reversed his initial enthusiastic support for spontaneous generation and by now had turned his influence as editor of *Monthly Microscopical Journal* and *Popular Science Review* fully to the cause of opposing Bastian. With Dallinger came the support of the Christian Evidence Society and the Wesley Scientific Society, organizations trying to make the new Darwinian science compatible with liberal religious teachings, and in this enterprise fully supporting the strategy of cutting off evolution from spontaneous generation claims.

William Roberts of Manchester had opposed Bastian since January 1873 on experimental grounds, and he now rose to Tyndall's support, adding an argument from evolutionary theory to his challenge. Writing in the *British Medical Journal* of 4 March 1876, he tried to deny Bastian's claim of offering the only view consistent with Darwin's theory. "The reluctance of some evolutionists to give up the spontaneous origin of bacteria," he urged

> is evidently due to the notion that this question is bound up with abiogenesis generally. This is a wholly erroneous idea. The question of abiogenesis will still remain after all have acquiesced in Pasteur's views of the origin of bacteria: indeed, to a logical evolutionist there would appear to be a strong *à priori* improbability in the abiogenetic origin of bacteria. They were not wanted, and could not exist, on the earth's surface until after other organisms had lived and died before them. Their special function and feeding ground lie amid the wreck of living things. And if the survival of the fittest hold good in regard to bacteria, they must be the remote progeny of less perfect organisms of the same class. What can be more perfect than their adaptability to their place and use in the order of Nature?[34]

Roberts added, in a note prophetic for the later course of origin of life research, "Those who are in search of a case of abiogenesis, should seek among the primitive organisms—if there be any such—which can exist and grow amid inorganic elements, in the water of the sea, or the mineralized springs and streams of the land."[35] This represented a line of argument that Tyndall himself never pursued, so certain was he that the issue at stake was Bastian's experimental incompetence.

The Exact Sciences versus the Biomedical Sciences

Tyndall claimed from his first lecture on the subject in 1870 that the analytic tools of the more exact physical and mathematical sciences would surely throw light where the medical sciences had failed to do so, and seems to have been surprised that medicine should take offense at such a view, let alone effectively oppose it. Farley has shown that the experimental evidence was indeed inconclusive, despite Tyndall's feelings to the contrary, until at least the end of 1876.[36] However, once the debate had become more polarized and Tyndall had felt himself to be the object of many offensive remarks, he seemed incapable of even imagining that anyone who agreed with Bastian could possibly objectively view his opponent's case as equal to or stronger than his own. That there could be public support for Bastian seemed to Tyndall only possible if a) Bastian was an incorrigible demagogue, concerned only with his own personal aggrandizement, and b) the public at large were hopelessly backward in their understanding of science. As he wrote in a letter to Pasteur:

> A taste for science has become general, and the public press must defer to that taste. Its contributors, however, have not as yet had the culture necessary to enable them to judge between good scientific work and bad. They lack the critical power which would enable them to perceive directly, and by internal evidence, the experimental weakness and logical inconclusiveness of Dr. Bastian; and the experimental strength and logical clarity of M. Pasteur. They are guided by knavery of assertion, and as Dr. Bastian is not wanting in this, they have leaned to his side. But education is advancing and this critical power will gradually come. Our public writers will become more and more able to distinguish between real and spurious teachers. To this end of public culture your works will powerfully contribute.[37]

Tyndall continued on this theme in another letter:

> A few minutes ago I received the British Medical Journal containing the translation of your article on Dr. Bastian's experiments. I forward it to you along with this letter. I also send you an English abstract of the paper presented by Dr. Bastian to the Royal Society. I send you further

> the abstract of your communication which was published in the Lancet. Also a letter from Dr. Bastian published in Nature Aug. 10. These will enable you to see the manner of man Dr. Bastian is. I shall deal with his references . . . in due time; I could not share the hope that you expressed regarding his abandonment of the doctrine of spontaneous generation—at least his open abandonment of it. Nobody could be more inclined to deal tenderly with Dr. Bastian than myself. But tenderness to him is sure to [be] misinterpreted and misused. His object is victory, not truth. He knows very well that in the present state of education he is sure to have followers if he is only bold enough in his assertions. He is however, simply postponing the inevitable day when his work will be reduced to its true value.[38]

A few days later, he wrote: "Bastian will fight to the last: we have got a habit in England of sympathizing with and praising *courage*. Men who are not courageous often take advantage of this. But the assumption of courage must be stripped from Dr. Bastian, and the English public must see him as he is."[39] What is noteworthy, again, is the accusation that Bastian did not worship, as Tyndall professed to, at the altar of Truth, but rather at the idolatrous altar of Victory. Tyndall's litany of accusations had moved well beyond experimental incompetence.

Another prong of Tyndall's public rhetorical strategy was his attempt to undermine the symbolic power of the microscope. This instrument was one of the most powerful icons of the medical and life scientists in their attempt to portray the physicist as an interloper in territories beyond his expertise.[40] Defenders of biology also used it to show that physics itself did not have the weapons to justify a hegemonic incursion into the territory of disease, as Tyndall claimed it did. A significant part of the case made in Tyndall's 1876 Royal Society paper was an attempt to refute arguments that Tyndall's "atmospheric germs" must not exist because no such things could be seen by the microscope, even in samples of some infusions that did become turbid and putrefy after boiling. Tyndall cited his own experiments and those of numerous others, showing that particles did exist that scattered light, yet were too small to be seen by the most powerful microscopes. He argued that his concentrated light beam would thus extend the power of detection of germs into a realm where light microscopes were inadequate. In a direct challenge to his opponents, he insisted that only physical science possessed the tools really adequate to rule in the germ-realm. His attackers, Tyndall said, had claimed that the inference of sub-microscopic germs was "the pure work of the imagination, resting . . . on no real basis of fact. But in the concentrated beam we possess what is virtually a new instrument, exceeding the microscope indefinitely in power. Directing it upon media which refuse to give the coarser instrument any information as to what they hold in suspension, these media declare themselves to be crowded with particles—not hypothetical, not potential, but actual and myriadfold in number—showing the microscopist that there is a world far beyond his range."[41]

In February 1876 Tyndall wrote enthusiastically to Pasteur: "I have gone over a great deal of the ground taken up by Dr. Bastian and have, I trust, refuted many of the errors by which the public was misled. The change which has occurred in the tone of the medical journals of England is very remarkable; and I am inclined to believe that the public faith generally in Dr. Bastian's accuracy has been considerably shaken . . . PS I do not think the germs of bacteria come within the range of the microscope."[42]

In one analogy, Tyndall likened patients in a hospital ward to just so many identical tubes of infusion into which disease germs might fall, importing the rhetoric of exact quantitative science by ignoring individual differences in resistance among patients. Tyndall argued that whether or not a given patient happened to develop disease depended entirely on whether or not a germ happened to fall into or near that "tube of fermentable broth." Those who did not must have been those on whom no germ happened to land. Adam has observed that Tyndall's arguments showed a "remarkable insensitivity to current medical knowledge and the crudeness of his [own] theoretical stance" toward it.[43] To a profession that had been actively debating the relations between "seed" and "soil" in infectious disease for hundreds of years, Tyndall's attempts at exact quantification (mapping of "germ clouds" and so on) appeared an unbelievably naive attempt to simply deny any role for the "soil," i.e., to ignore one of the most significant facts of medicine: the "constitution" of the patient.[44]

His claim of dominance for the "searching beam" was viewed with similar skepticism due to lack of experimental support. Even Pasteur never picked up on Tyndall's suggestion, since he plainly believed that bacteria themselves, and other objects in that size range, were the "germs," whereas Tyndall continued to speak of the "germs of bacteria"—their much smaller spores or seeds. Lionel Beale had already attacked Tyndall on this point in private. In response to this new, more detailed challenge Beale, the premier microscopist of the day, sharply criticized Tyndall's method as unable to show the difference between organic and inorganic particles or even to distinguish between dead and live bacteria when these were present. Since the microscope could do these things, it was not surprising, Beale argued, that the method of Tyndall had been put to little use.[45]

Among the doctors, even William Roberts, who had opposed Bastian from early on and was not known for being as irascible as Beale, found himself in agreement with both of them that "[t]he ingenious attempts of Pasteur and others to demonstrate germs in the air are manifestly illusory. Like them I have repeatedly collected air-dust and found abundance of molecules, circles, spheres, and particles of various kinds under the microscope; but these could not be identified as true spores, nor distinguished from particles of inert dust."[46] Unlike many of his medical contemporaries, however, Roberts was willing to conclude from this, like Tyndall, that the microscope itself must be at fault, since he deeply believed that germs exist.

That Tyndall did not take kindly to Beale's statements about him is evident in a letter to Huxley shortly afterwards in which he says: "I think I must ask you one of these days to have a close look at my infusions. An authoritative word from you will sweep the nonsense of that little humbug Beale from the face of the earth."[47] Huxley, for his part, participated in Tyndall's experiments and reviewed his draft papers, trying to persuade his friend to be a bit less rigid in his zeal. On one draft, he commented: "The other point worth considering is the very absolute way in which you ascribe your burrowing abscess [from Tyndall's 1869 accident in the Alps] to Bacteria. No doubt Bacteria excite suppuration when they get into a wound, but no doubt either that suppuration may take place to any extent without their agency. I think you lay yourself open to attack by the doctors as the matter stands."[48] Huxley, despite his opposition to spontaneous generation, still seemed much less absolute in his stance on the germ theory of disease. And he clearly had a tactical eye on keeping "the doctors" among evolutionary supporters in line.

The X Club and the Royal Society

Despite the still outspoken deportment of "the doctors," X Club influence in the Royal Society had steadily increased through the 1870s,[49] and in 1876 this clearly had an impact on the confrontation developing there between Tyndall and Bastian. Tyndall delivered his first major attack at the Royal Society, with Hooker as president in the chair, on 13 January 1876. Bastian was not intimidated and arranged to meet with Tyndall to compare experiments. Fully confident of his own experiments, Bastian remarked to his friend Norman Lockyer, "We are to have an interview . . . on Friday next. He [Tyndall] to show me some of his results & in turn to look at mine. This was not accorded without some hesitation—and I hope he won't draw his head out of the noose too soon!"[50]

Bastian then took on the arguments of Tyndall's paper in a lengthy paper of his own submitted in May. But while Tyndall's paper had been quickly praised by referees and recommended for publication in the prestigious *Philosophical Transactions*, a decision on Bastian's paper stalled. Alexander Williamson, chemistry professor at University College, and Huxley were assigned to referee Tyndall's paper, and both being close collaborators of his, could hardly have been expected to be unsupportive. Both gave their positive reports within a few days of being assigned the task, and the paper, having been received on 6 April, appeared in print in the *Philosophical Transactions* by 28 July, unusually speedy treatment.[51] Bastian's paper was received on 24 May, read on 15 June, and assigned for review in late July, i.e., only after a month delay, to Michael Foster and to John Burdon Sanderson. Neither man ever submitted a report, however, for reasons unknown. So the paper was again farmed out for review on 11 August to chemist Henry C. Sorby and, not until 28 October, to Dr. George Rolleston, physiologist at Oxford. Both men delayed giving their reports until January 1877. Meanwhile, Tyndall and Dr. William Roberts of Manchester each submitted another paper opposing Bastian on 18 December, both of which were read within three days, reviewed, and voted on for the *Philosophical Transactions* within a month.[52] While it was not unusual for some reviewers to take more than a month to submit their report on a paper, the disparity between the treatment given to Tyndall and Bastian simultaneously during this period is striking. The choice of Huxley to review Tyndall seems questionable, given Huxley's public antagonism toward Bastian, as does the repeated choice of Huxley lieutenants to review Bastian. However, given the fact that the Royal Society was a "small town" to begin with, as the younger generation more and more filled in with students of X Club members, it became increasingly difficult for Bastian to get a fair hearing in that forum.[53] Huxley's scorn for Bastian is plain in one letter to Tyndall during this time: "Bastian has been blundering again as usual. There is an action in the Scotch courts 'for putting to silence' a frivolous litigant. I begin to wish we had something of the same sort for the irrepressible B."[54]

In the end, Roberts's paper opposing Bastian, Tyndall's second attack, and Bastian's first reply were all voted on by the Society's Committee of Papers on 11 January 1877. Once all the reviewers' reports were in, Huxley could cheerfully tell Tyndall: "I shall be curious to see what line Bastian takes. The fates are hard upon him just now, as the referee reported dead against his paper to the meeting of Council last week, so it will not be printed."[55] Upon which Tyndall immediately crowed to Pasteur:

> You know the habit of the Royal Society is to place every paper presented for the Philosophical Transactions in the hands of two referees, who are chosen for their acquaintance with the subject. They report to the Council favorably or unfavorably as the case may be. This caution is not at all observed with regard to the Proceedings where many things of doubtful value are permitted to appear. Well, Bastian's paper was intended for the Transactions, an abstract of it only appearing in the Proceedings. Both the referees have reported dead against it; so that it will not be permitted to enter the Transactions. I communicate this to you confidentially. It will soon however be generally known.[56]

Bastian at first protested to Huxley, asking the reason why the Commission of Papers directed his last paper "to be deposited in the Archives of the Society" and asking for the name of the referee.[57] Huxley replied with a formality that contrasts sharply with his tone to Tyndall. He stated simply that names of referees were always treated as confidential, and offered to lay Bastian's letter before the Council if he so desired.[58] Of course, Bastian, still smarting from the debacle that followed Huxley's similar offer of three and a half years previously, realized that this would do him no good and that the Royal Society as a forum for his views had now been effectively denied him by the X Club. His next (and final) major experimental paper directed at Tyndall was delivered instead to the Linnean Society a few months later and published in that society's journal.[59] In the meantime his 1876 ms. was now so out of date that he chose to abandon it, and it remains to this day in the Royal Society Archives where few, surely, have ever read it.[60]

Meanwhile, Tyndall delivered his victory lecture in May 1877.[61] By this point many observers on the sidelines, including Darwin, had been convinced. As Darwin wrote in response to the latest news from one who was present: "I had heard nothing of the 'rub' at the Royal Society . . . I suppose it refers to 'spontaneous generation,' and I shall be glad of anything which helps to settle the question for the present. Huxley recently told me that he thought Tyndall's recent work about old germs withstanding long-continued boiling was most important and apparently decisive."[62]

Spores: Tough Allies to Kill

In addition to recruiting people and institutions, Tyndall also recruited objects with strong symbolic content. In addition to J. B. Dancer's "dust" (which carried the emblematic weight that "filth" did for his opponents in the medical camp),[63] and Pasteur's silkworms and "atmospheric germs," Tyndall brought into his stable the "dust-free chamber" (see figure), the bearer of pure, limpid, sweet-smelling broths by virtue of being proof against the germ demons. Tyndall had designed this sealed cabinet soon after his first clash with Bastian in 1870. The inside was coated with glycerine (a trick he picked up from Pouchet)[64] to trap all dust particles once they settled out of the air. The glass windows allowed a light beam to be shined through, so that the observer could verify when absolutely no dust remained suspended prior to beginning an experiment. Tyndall's claim was that tubes of infusion boiled and then allowed to stand open in this dust-free environment never showed any turbidity or putrefaction, even after standing for months; but once exposed to normal, impure air, they would usually show signs of bacterial growth within a day or so at most. Tyndall argued that this proved, plain and simple, just like Pasteur's swan-necked flasks, that atmospheric dust must contain the germs of bacteria and other organisms found in the "contaminated" infusions.

Bastian countered that Tyndall, like Pasteur before him, had *at most* proven only that dust was a necessary ingredient for archebiosis or heterogenesis in some circumstances. In other experiments, where the physicist sealed his tubes in a flame (or had his proxies do so) to allow greater portability, the result was more equivocal: "tubes containing his infusions would, *for the most part,* keep pure as long as they were sealed, but when exposed to air they grew germs."[65]

Bastian also did many experiments in which well-sealed tubes, especially of hay and turnip-cheese infusion, were boiled, sometimes for extended periods of time, and yet showed prolific growth of microorganisms after cooling. Tyndall insisted that all this showed was that Bastian had not boiled his tubes long enough, or at a high enough temperature to kill the "germs," or perhaps that he was just careless when it came to ensuring that the tubes were totally sealed. Others such as the German bacteriologist Ferdinand Cohn even suggested that the microorganisms might produce heat-resistant spores, analogous to the desiccation-resis-

Tyndall's dust-free chamber for spontaneous generation infusion experiments, 1875–1877. Courtesy of the Royal Institution.

tant forms known for many tiny organisms to be amazingly resistant. Given the lack of *proof* for any such spores, however, Bastian and other critics attacked this as an *ad hoc* hypothesis. Clearly both sides were arguing from theoretical suppositions. Since at that time there was no agreed-upon consensus as to the conditions required to sterilize an infusion, those who agreed with Bastian that no living protoplasm could withstand the boiling temperature of water for more than a few minutes found the assumption that microbes could be generated *de novo* less *ad hoc*. In Tyndall's worldview, for example, the analogy of growth of organisms from seeds held a far higher and more inviolable place than the notion of organisms *somehow* being able to withstand boiling water. The result was a classic example of what Harry Collins has termed "the experimenter's regress."

Tyndall's discovery of Cohn's heat-resistant spores in late 1876 enabled him from that point forward to reconstruct a persuasive history of the debate to explain that all previous confusion could be attributed to the presence of such spores in the infusions of Bastian and others. As tempers cooled by the 1880s, Cohn's spores allowed Tyndall to abandon the kind of shrill moralistic demonizing of Bastian that characterized the most bitter exchanges of 1876–1877. A less morally loaded claim, it did not even have to mean that Bastian was a poor experimentalist. This point deserves great emphasis: even before the discovery of spores (a process itself significantly driven by the interest in disproving Bastian),[66] Pasteur and Tyndall were nonetheless sure that their opponents were wrong, begging the question to at least the same extent to which they accused Pouchet and Bastian of doing.[67] Yet even the few historical treatments to date that have acknowledged this have still stayed with a primarily *experimental* narrative of what the debate was about. Prior to the discovery of spores by Cohn and Koch, Tyndall played down the occasions when a tube of his that was not exposed to dust ever "became contaminated."[68] By the height of his personal campaign against Bastian in early 1876, he was declaring with much fanfare that tubes of infusion boiled for a short time and then protected from dust would *never* yield microbes, "though given five hundred chances." His level of invective implied that whenever Bastian's tubes showed growth, experimental incompetence on the younger man's part was the likely cause: "The evidence . . . that Dr. Bastian must have permitted errors either of preparation or observation to invade his work, is, it is submitted, very strong."[69]

In cases where dust getting into a tube was shown to be related to the growth of microorganisms, Tyndall was even more close-minded: by analogy with plants coming from seeds, he stated, "It would be simply monstrous to conclude that [microbes] had been spontaneously generated."[70] For Tyndall, then, "purity" and "contamination" had been just as much moral descriptions of his and Bastian's motives as they were descriptions of the state of microbial growth in tubes of boiled infusion.[71]

However, Tyndall's insistence that *only* protection from dust was necessary to guarantee sterility on all occasions had committed him to a position that was difficult to reconcile with the discovery of heat-resistant spores. Farley points out the "obvious betrayal of the 'scientific method'" inherent in Tyndall's arguments at this point, and the irony that, despite this, "Tyndall later had the gall to attack Bastian for being 'unacquainted with the real basis of scientific inference.'"[72] He points out that Tyndall had to make an embarrassing public turnaround, if he wished after October 1876 to use his friend Cohn's spores as the explanation of his current problems.[73] This Tyndall did, again characteristically, with much fanfare (about all the extra work required); however, this left a somewhat awkward theoretical break between his 1876 and 1877 papers on the subject. To set the stage for his new theoretical stance, Tyndall announced at the Royal Society on 18 January 1877 some surprises contradicting his previous work:

> Cucumber infusion has been subjected, for intervals varying from five minutes to five hours and a half, to the boiling temperature without losing its power of developing life . . . I tried to reproduce the results with animal infusions obtained with such ease and certainty a year ago . . . But in my recent experiments, where the care bestowed far exceeded that found necessary in my last inquiry, the animal infusions, like the vegetable ones, fell, for the most part, into putrefaction . . . I have also pursued my experiments with closed chambers, from which the floating matter was removed . . . Precautions far greater than those found successful a year ago failed to protect these infusions from contamination.[74]

The sharp discontinuity was glossed over by the audience, and Tyndall's rather *deus ex machina* explanation of a bundle of hay having contaminated the previously pristine Royal Institution was lauded rather than skeptically scrutinized. This portrayal alone could have appeared highly suspect because of Tyndall's previous discussion of how tubes left open in all parts of the building, including the cellar and the roof, all putrefied. The busy traffic of Albemarle Street must have made this a very dusty environment in which to work![75] More seriously, however, it is possible that Tyndall only initiated his new set of experiments *in response to* the discovery of heat-resistant spores. His laboratory notes on this series begin 7 October 1876, just days after first reading Cohn's journal. This suggests that Tyndall may have begun these experiments in a frantic effort to find out if he may have been wrong in assuming up to this point that any growth in a boiled and well-sealed infusion must mean error and contamination, such that he had been unwilling to admit that this result ever happened among his own tubes.[76] Tyndall's use of spores, like Huxley's earlier annexation of "Brownian movement" to be turned against Bastian, can be much better understood in the context I have just sketched: the opposing claims of Tyndall and Bastian are in the two papers for whom the Royal Society was the audience, that same audience that had shown itself so sympathetic to Tyndall and treated the two men's papers so nonsymmetrically.

Bastian's response, which the Royal Society would likely not have published but which the Linnean Society did, was justifiably scornful at Tyndall's sudden change in the thrust of his argument:

> what Prof. Tyndall had been unable to achieve in the way of inducing fermentation in boiled and guarded fluids, had three years previously been brought about by me in the presence of a highly skilled and then skeptical witness [Burdon Sanderson] . . . Meanwhile, almost at the same time that the learned physicist was acting in this bewildering manner, . . . Prof. Ferdinand Cohn, was again confirming my impugned experiments . . . and was obtaining . . . those evidences of fermentation which hitherto Prof. Tyndall had strangely enough failed to reproduce. The fact was again fully admitted by Prof. Cohn, though my interpretation of it was still questioned . . . Twelve months later, we find Prof. Tyndall announcing that he was then able to obtain the previously denied results. The behaviour of his recent infusions had completely stultified his previous position. He was no longer at issue with me and others in regard to the fact. The difference between us was now one of interpretation only. In spite of his previously much-vaunted 500 negative results, and the good evidence which they supplied as to the death-point of Bacteria and their germs, Prof. Tyndall

now endeavoured, as best as he could, to cover his previous unfortunate position. The result was a complete change of front.[77]

Convincing Pasteur: Urine Proves a Weak Ally

In the meantime, however, Tyndall continued hammering away to convince Pasteur that, regardless of the evidence, Bastian was not just mistaken but a shameless opportunist. On several occasions he persuaded the illustrious Frenchman to write letters to British journals in response to articles by Bastian. Pasteur complained at one point that some of Tyndall's communications were too polemical and harsh.[78] Tyndall responded by trying to keep the focus on his image of Bastian as a terribly unskilled experimenter (spores had only just come to his notice at the time of this letter and had not yet affected his rhetoric):

> I do not know whether you have read Dr. Bastian's two volumes entitled the 'Beginnings of Life,' and a third volume called 'Evolution or the Origin of Life.' [sic] I have carefully read these and venture to think them the worst specimens of experimental work that ever came under my notice. I am perfectly aware of Dr. Bastian's cleverness. Were he not clever, he could not have influenced the world as he has done. He is now beginning to see the grossness of his own blundering; this renders him far more cautious, and his last paper was written under the influence of this caution. I think caution will be required on your part in dealing with him, for he is sure to convert any expression of approval on your part to purposes which you would never think of applying it. Trust me when I say that he fights for victory and not for truth. I have made scores of experiments upon the neutralized urine, but they are far from justifying the position of Dr. Bastian.[79]

The last line refers to a new offensive front opened by Bastian during the summer of 1876. He announced that a new experimental system could produce growth of microorganisms where, according to Pasteur, none should occur. Bastian declared that sterile urine, whose acidity was neutralized by addition of solid potash (KOH), would show proliferation of microbes after a few days. Pasteur had reported results that confirmed those of Bastian, though disagreeing with the interpretation.[80] Tyndall was quite concerned that Pasteur should insist upon his differences with Bastian, and Pasteur agreed to write a letter to the *British*

Medical Journal. Tyndall, still worried that Pasteur's tone was too generous, wrote back:

> I am much interested in what you tell me regarding your experiments. I cannot, however, understand your corroboration of Dr. Bastian's experiments, as announced in the Comptes Rendus of July 17th . . . I hope you will be careful in dealing with Dr. Bastian. Give him strict justice, but any generosity you show him he is sure to misuse. No man could be more disposed to be generous to Dr. Bastian than I was, but his conduct renders generous treatment impossible. He still succeeds in mystifying a large proportion of the medical profession.[81]

Pasteur sent Tyndall a copy of his next "reply to Bastian" to be published in the *Comptes Rendus* of the French Academy, to which Tyndall responded:

> Your note is excellent. Nothing could be better; and it has arrived at a very opportune moment . . . You are quite right in postponing the publication of your note. It will produce all the better effect, afterwards . . . Before I publish, I will send you a copy of what I propose to publish. The mass of men are so unenlightened and stupid, and so fundamentally stupid in relation to this question, that any difference between you and me will be sure to be misinterpreted; or at all events to receive an interpretation extending far beyond the real limits of this difference.[82]

It is not clear whether Pasteur shared Tyndall's cynical assessment of the public's understanding of science. However, Pasteur would still not declare Bastian incompetent or ill-intentioned, as his self-appointed British mouthpiece urged. In a letter a few weeks later Pasteur enclosed a copy of a recent letter he had written to Bastian, in which he expressed respect for Bastian, in spite of their differences.[83] Soon after, Pasteur noted having received a very polite reply from Bastian.[84] Tyndall, fueled by the Royal Society's rejection of Bastian's paper at just this time, replied:

> I thank you very much for the glimpse of Bastian's letter. It is very useful to me to know his present notions . . . His letter is perfectly characteristic. He will not do what you require him to do; but will shift his ground if you permit him to do so. Your praise of his experiments is sure to be turned to a mischievous account. You see how he endeavors to make it appear that you and I oppose each other. I find some of his

> experiments utterly wrong. You find a totally different set of experiments right, and on this he founds the conclusion, *and will succeed in spreading it abroad,* that you and I are utterly opposed to each other. I return his letter with many thanks.
>
> PS I say with deliberation that taking the whole of it into account, Bastian's experimental work is the *worst* that I have ever known.[85]

In his own letters to Bastian, Pasteur had hoped that Bastian would soon come to see that his conclusions were in error, and to accept correction from a much older and more distinguished man who had taken an interest in his work. Over time, as Bastian held his ground and seemed no more impressed by Pasteur's credentials than he had been by Tyndall's, the chemist became less patient. By January 1877 Pasteur was urging the younger man a bit more stridently to recant: "I entreat you, in the name of truth, to confess loyally that the conclusions you have advanced on the subject of the spontaneous generation of bacteria in the solution of urine and potash are entirely erroneous. You will gain by doing so a reputation for scientific probity and honor, which will add more to your name and to the distinction of your career as a conscientious scientist than even an important discovery."[86]

Tyndall kept up the pressure, angrily reporting the latest support for Bastian in the 17 February *Lancet.* "The *Lancet* is the leading medical journal of England. It has a very large circulation, so that an error promulgated by it is likely to do considerable mischief. I think this error . . . ought to be corrected. Dr. Bastian at the meeting of the Royal Society last night referred to the high terms in which you have spoken of his work. He parades your testimony regarding the correctness of his experiments on every possible occasion."[87] Pasteur reported that he had found the source of error in Bastian's urine experiment: that the microbes would not appear in the solution if it was neutralized with boiled KOH solution instead of solid potash. As late as 24 February 1877, he still wrote to Tyndall that Bastian was in error, but had made his mistakes in good faith.[88]

Bastian refused to accept Pasteur's claim that this meant the growth in his tubes of urine must be due to germs contained in the solid potash. He challenged Pasteur in the *Comptes Rendus* of the French Academy to submit both sets of experiments to an independent scientific commission, which would rule on whether experimental error or lack of care was to be found in his technique, settling Tyndall's endless accusations once and for all. For the French chemist, this brash intrusion into his own institutional territory seems to have been the final straw which turned him toward a strategy more in line with Tyndall's recommendations. Pasteur had as much confidence in firm support from the French Academy as Tyndall and Huxley had in the Royal Society, and he wrote: "I should have willingly spared Dr. Bastian the condemnation of an academical commission," however, he "has persisted in shutting his eyes against the truth."[89] Tyndall was elated to see the confrontation developing. Pasteur wrote to tell him of the formation of the Commission, and of how popular Tyndall's work was becoming in France. Tyndall replied, "I was pleased . . . to receive your last note . . . of your discussion with Bastian. Care will be needed in dealing with him; but you must have felt the necessity of this. If he can gain victory on a side-issue, he will try to do so."[90] Though Pasteur's invective against Bastian never quite reached the level of Tyndall's campaign, from this point on the immense reputation of Pasteur was effectively at the service of Tyndall's rhetoric and not accessible to Bastian's most skilled tactics.

It is noteworthy that no one experimental "blow" was ever decisive in resolving the issues at stake, nor was any one exchange ever considered to have gone entirely to either Bastian or Tyndall. The medical community, in particular, continued to feel skeptical of the germ theory and unconvinced that Bastian had been conclusively defeated for some years afterwards. Rather, eventually the audience lost interest in each exchange when it perceived stalemate to have been reached. The debate was revived when the stage instead shifted to new experimental setups and new institutional forums. Denying Bastian some of the major forums, by behind-the-scenes maneuvering, gradually diminished his access to audiences. This in turn precluded him from continuing to be perceived as a respectable representative of evolutionary science. The considerable damage to his personal and professional standing that followed from Tyndall's actions would have further curtailed Bastian's access to students among the young generation of scientists in training. Though there was still considerable disciplinary support within medical circles, as well as theoretical opposition to a simplistic germ theory of the type Tyndall promoted, Bastian largely withdrew from public engagement in the debate by late 1878. He no doubt hoped thereby to induce Tyndall to cease the campaign to destroy his reputation as a scientist, particularly because he was a candidate for a full professorship at University College (which he succeeded in obtaining) in that year.[91] It is clear that his opinions on the subject never changed, however, and when he retired twenty years later as a world-famous neurologist, he again actively commenced laboratory work and publication to make the case to the scientific world for the reality of heterogenesis and archebiosis. That story will be taken up briefly in the next chapter.

With regard to Tyndall's energetic campaign against Bastian, this story demonstrates that it was deeply rooted in the larger X Club campaign to win for science a voice of cultural authority. My focus on the issue of the experimenter's regress and the underdetermination of the issue by experimental evidence does not deny that experiment was a very important part of the spontaneous generation debates in Britain. It has, I hope, finally laid to rest the use of this episode as a "textbook example" of where "nature spoke, and *that alone* resolved the controversy."

Conclusions

Adrian Desmond's work has exploded the entrenched view that spontaneous generation never had any serious audience in the natural theology context of Britain. He has shown that in the 1830s and 1840s a very important audience did exist, namely, radical medical agitators. In the preceding chapters I have shown that the audience for spontaneous generation had made significant headway by the 1860s into much more respectable Victorian scientific circles. Owen, Bennett, and those doctors who vigorously opposed the germ theory were its main supporters. The fact that these were "losers" in other areas explains in part why the story of any significant support for spontaneous generation in Britain has been so effectively lost.

In a famous article William Coleman argued that cellular continuity from generation to generation was a crucial requirement for Darwin's theory of natural selection.[1] If this were true, it should be logically impossible to fit new cells generated spontaneously without mitosis into a Darwinian framework. In line with this, previous portrayals have tended to paint Bastian as an aberrant thinker in the British context, more like some displaced 1850s German materialist than a British "scientific naturalist" of the 1860s and 1870s. But many Darwinians did not accept cellular continuity in the 1860s and 1870s, and Darwin himself waffled on the spontaneous generation issue during those years. The "scientific naturalist" category, first developed by Frank Turner and L. S. Jacyna, has great analytic usefulness; however, all the primary beliefs constitutive of scientific naturalism—conservation of energy, evolutionism, and atomic theory—are precisely those features in which Bastian's theory of spontaneous generation was completely up to date and in agreement with Darwinian evolution and Huxleyan physicalism. Nor did Bastian merely use those principles so that his views on spontaneous generation could become more scientifically acceptable. Bastian and many British supporters of evolution felt that his reasoning on spontaneous generation was a logical development of evolution and the theory of conservation of energy, even before he had done any experimental work on the subject. The widely perceived connection with evolutionary theory, then, was the most important pillar of support for spontaneous generation.

But we must also consider the flip side of this coin. Alison Adam points out, "Huxley 'saw' *Bathybius* at the height of his enthusiasm and support for the protoplasmic theory of life," i.e., "when theoretical considerations were optimal for interpretation of the mud samples as an organism made up only of protoplasm."[2] She concludes, sensibly, that "Huxley must have been quite glad to relinquish *Bathybius* in the end as Haeckel's enthusiastic linkage of the organism with abiogenesis put him in an embarrassing position." But on the British scene, the debate was not mostly about Haeckel. Once we acknowledge the importance of theory-laden observation, we also ought to ask: Is it totally coincidental that Huxley changed his mind and chose to "see" *Bathybius* as an artifact in 1875, just at the height of his desire to distance himself (and evolution) from Bastian?

Lionel Beale, seen by many as an extremist who criticized just about everybody, was not the only person to view Bastian as a member of the naturalist camp, along with Huxley and Tyndall. Since Huxley's and Tyndall's writings repeatedly make this point, ignoring Beale's criticisms, or writing them out of their histories, was effective for Huxley and Tyndall in creating the historical impression that Bastian was just as far removed from them and just as "wild" as Beale. Lost in the process is the fact that many thought Bastian a cutting-edge experimentalist and thinker, and that among these were supporters of Tyndall, Huxley, evolution, and naturalism, not just their opponents.

Thus, in addition to his experiments, it was because of Bastian's *success* as an orator and rhetorician and because of the widespread belief in Britain, among both the allies *and* opponents of Darwin and Huxley, that spontaneous generation was logically implicit in naturalism, that Bastian's theories were taken so seriously in Britain through most of the 1870s. It was not solely or primarily theoretical incompatibility that led the X Club to oppose spontaneous generation; indeed, Huxley, Spencer, and Frankland took a lively interest at first in whether Bastian might prove spontaneous generation possible. Thus, to invoke scientific naturalism as an explanation of the X Club's opposition to spontaneous generation merely begs the question. Instead, the X Club ultimately rejected Bastian because of their overwhelming fear that his work might lead to a repeat of the *Vestiges* and Crosse episodes, which "had made philosophical naturalism look ridiculous and free-thinkers gullible"[3] in the British scientific world. Tyndall was further motivated as early as December 1869 by a specific commitment to the germ theory of disease, convinced that only by its acceptance would medicine become scientific.

I am not arguing that the X Club was the only significant force in Royal Society politics or on national science policy. Clearly, a lot of wheeling and dealing went on in a body like the Royal Society of London, and Huxley and his colleagues were expert players. Then too, many members may have been uninterested in the political maneuverings or even avoided taking part in them. The complex workings of psychological and ideological forces within institutions cannot be reduced to simple power games, even power games motivated by what the participants actually believed to be the best science. However, to leave out the power asymmetry as one dimension of the spontaneous generation story, as previous histories have, also distorts the picture of what occurred.

Furthermore, if the power dimension is ignored, our attention is distracted from an interesting process: Huxley's communication to young scientists of standards of behavior appropriate to their trade. Learning the appropriate rules of conduct is just as much a part of the acculturation of trainees as learning the practices needed at the laboratory bench or in the field. Steve Shapin, Simon Schaffer, and Robert Kohler have described how these standards are communicated and enforced. Their work also shows that proper conduct is often a crucial factor in the outcome of a debate.[4] Having Huxley, Sharpey, Busk, and Frankland witness his early experiments gave Bastian's early work a boost in credibility. Bastian benefited again from this validation by proxy when Burdon Sanderson published confirmation of his technique and observations in January 1873. However, the double-edged nature of witnesses became clear when Huxley, in September and October 1870, claimed that by witnessing Bastian's work he could authoritatively declare it to be based on faulty technique. Still, Huxley had not been Bastian's only source of support, and many scientific onlookers throughout the conflict from 1870 to 1878 stated that experiments on both sides seemed free of fault. In those circumstances they repeatedly demurred any conclusion, since two men of unimpeachable scientific reputation were involved. Given the importance of reputation in scientific debates, it is not surprising that Tyndall's final victory over Bastian depended upon his active campaign to undermine Bastian's reputation as well as the validity of his experiments.

Ruth Barton has noted how completely the accounts of the development of evolutionary science have been dominated by the versions of that story constructed by Huxley, Hooker, Tyndall, and others. This has operated right down to the master narrative of which journals actually carried the message of the "real" evolutionists, beginning with the *Natural History Review,* the *Reader,* and finally *Nature.*[5] Restoring the role of more popular science journals is thus a necessary part of reconstructing the full story of the spontaneous generation debate. Lawson's *Popular Science Review, Monthly Microscopical Journal, Scientific Opinion,* and, to a lesser extent, the *Journal of the Quekett Microscopical Club* reached much wider audiences than the more technical and expensive journals such as the *Quarterly Journal of Microscopical Science,* and the scientists read them as well. Lawson supported heterogenesis, Pouchet, Bennett, and Bastian early on, and he used editorials, book reviews, and reprinted articles in his journals to promote that cause. The X Club does not mention Lawson and his journals, probably because they represent so graphically that a faction of evolutionists supporting Bastian did exist. Lawson's shift against Bastian is also a significant barometer of the dispute. After that shift, his journals were every bit as actively used as they had previously been—but for opposing spontaneous generation, especially via promoting the work of Dallinger and Drysdale.

Thus, Bastian, though perhaps not representative of the majority in the extent to which he had moved beyond natural theology, was not an aberration, but a quintessentially British phenomenon. He dared to out-Huxley Huxley and to draw conclusions from Tyndall's statements on the conservation of energy that showed a contradiction with Tyndall's own commitment to the germ theory. In addition to experiments, it was a series of historically contingent social and political losses by Bastian's allies, and successes by his opponents (and even Huxley's opposition was contingent, not becoming firm until April or May 1870), that led to the domination of the Tyndall/germ-theory version of scientific naturalism rather than the Bastian/anti-germ-theory version of scientific naturalism. The derailing of Bastian's scientific reputation, especially as an evolutionist, therefore requires explanation just as much as the derailing of Richard Owen's: Bastian was not generated *de novo,* nor did he easily disappear as soon as anyone tried to challenge him experimentally. Just as Farley and Geison have shown that experimental evidence was not sufficient in 1864 to account for Pouchet's demise at the hands of Pasteur, I would argue that other factors were needed to sink spontaneous generation in Britain. Experimental evidence was one of these factors, but internal contradictions in Tyndall's experiments made this insufficient. Pasteur's student Emil Duclaux came closest to an accurate assessment of Bastian's significance in the history of the debate (much more so than any British writer retrospectively describing it) when he said that Bastian's experimental proficiency and rhetorical staying-power were responsible for forcing the opponents of spontaneous generation to see the previous errors in their own work, and to discover the heat-resistant bacterial spores that were responsible, much sooner than might otherwise have occurred.[6] The quick retraction of Huxley's and Tyndall's claims that Bastian's work was experimentally incompetent shows how premature their certainty was in September 1870 that Bastian "got out of his tubes precisely what he put into them." Only a clear message from the powerful X Club to the young rising evolutionists, culminating in an energetic campaign by Tyndall to destroy Bastian's scientific reputation, was finally sufficient to overcome the attraction many felt to Bastian's combination of experimental skill and rhetorical power.

Further, although Bruno Latour has shown that one of the most crucial forces behind the success of Pasteur's germs in France was the fact that a previously stymied and ineffective sanitarian movement saw those germs as helpful allies, in Britain the reverse was true. The sanitarians had been spectacularly successful already by 1870, and had done so with Liebigian "zymotic ferments" as their allies (if we wish to use Latour's language), so that germs were not only unnecessary but were actually competitors, trying to steal the glory for a victory hard-won by the labors of the chemical poisons.[7]

The lesson is not, then, just about how "errors" (such as Bastian's lack of knowledge about heat-resistant spores) are eventually "found to be wrong." The deeper issue is how thoroughly the process of *defining* which things are "errors" and "artifacts" (such as Huxley eventually admitted *Bathybius* to be) depends upon a complex social process of negotiation among the competing claimants. Often these negotiations involve the nonexplicit use of broader cultural images as standards against which the competing theories are being judged. The *Bathybius* case showed that serious and experienced microscopists, beginning with Huxley, could look at globs of an artifactual chemical precipitate from ocean sediments preserved in alcohol, and see those globs as primitive living things, because new physicalist theories called for such primitive unorganized protoplasm. Haeckel's phylogenetic reasoning on the role that primitive "Monera" should play in filling the gap between the simplest cells and the most complex albuminous organic compounds provided the mental niche into which the globs could slither. (Although inanimate, they were seen to show lifelike movements, reported by well-established men like physiologist William B. Carpenter, as the publicity and enthusiasm over *Bathybius* grew.) At the time, some microscopists were quite critical about the rapidity with which such claims were being made. George C. Wallich was one, although he was frank enough to also admit a role for his own Victorian cultural preconceptions:

> Although new facts may dawn upon us, and new triumphs of mechanical and optical skill may hereafter enable us to detect subtleties of structure as yet invisible to our senses, . . . every new fact and every additional means of observation we may in future command will only serve to prove more incontestably . . . that even in her subtlest workings Nature still abides by Law, and permits no exceptional case—such as that which has been assumed to take place in these lowest forms of life—to disturb her harmony.[8]

Here Wallich shows that a Victorian concern that nature be law-abiding was such common cultural coin that it could be used as a criterion by which to scientifically judge interpretations of what was "seen" microscopically. What was at stake was *which* view of nature was most law-abiding. Bastian and others emphasized that an arbitrary break between the laws that govern physical forces and those governing vital forces was the most disturbing break of all in the "continuity of natural law."

Despite Duclaux's more sophisticated assessment of Bastian, the notion of "discovery" has such appeal as a pedagogical device that the vast majority of biology and microbiology textbooks have preferred Huxley and Tyndall's version of a Bastian who "with unusual determination, . . . held to his faith in spite of the fact that his arguments were destroyed with monotonous regularity by Pasteur and his collaborators."[9]

Jan Sapp's discussion of Franz Moewus is extremely illuminating here, especially with regard to "losers" such as pleomorphism, spontaneous generation, and Bastian.[10] Sapp argues that "fraud" (the term Tyndall applied to Bastian) is not defined in science as the mirror image of "truth," but rather is defined during the complex negotiations over a "discovery." In the process of defining a new "discovery," such as bacterial spores that could survive boiling, or cellular genetic continuity of distinct bacterial species, "fraud" became defined as the mirror image of the discovery. Numerous writers had advocated the *possibility* of heat-resistant spores or inferred that they had been "discovered" because spontaneous generation must be impossible. Bastian and those who thought it unlikely that life could survive the boiling temperature did not accept this as a "discovery." The "discovery" of heat-resistant spores finally achieved general assent only when it was reified by "fractional sterilization" in the context of Bastian's severely eroded social position in the Royal Society. Thus, just as Rudwick has shown that Murchison could only retrospectively claim to have "discovered" the Devonian strata, for Tyndall to claim that the discovery of heat-resistant spores ended the spontaneous generation debate is a highly simplistic description.[11] And now that the understanding of horizontal gene transfer has drastically altered our sense of what "genetic continuity of a bacterial species" means, what will become of that "discovery"?

The X Club Darwinians essentially had become the London scientific establishment after 1870. This gave them control over Bastian's access to at least some important platforms for presentation of his views. This strategy was to prove more and more effective through the mid-1870s, until by 1878, well before the acceptance of the germ theory of disease among a majority in science or medicine, Bastian's reputation was so damaged that he was essentially forced to withdraw from public debate on spontaneous generation. He remained an important figure on the London medical scene, being promoted to full professor at University College Medical School in 1878, and continuing to publish respected work in neurology for two full decades more. But Huxley and Tyndall effectively destroyed the linkage between spontaneous generation and evolution. When Huxley and his students wrote the next generation of biology textbooks, they wrote as if Bastian and that linkage had never existed. The success of their version of the story can be gauged by the amount of historical work required to reconstruct the fact that a camp of Darwinian spontaneous generation advocates who not only *existed* but were a serious intellectual force in the 1870s, as different "Darwinisms" struggled among themselves for the survival of the fittest.[12]

The in-fighting among Darwinian factions over spontaneous generation significantly shaped the discourse on this subject in biology. Most important, once and for all the debate forced acceptance of a linkage between evolution and some kind of naturalistic explanation of the origin of life. However, Huxley's new term "abiogenesis" encapsulated a new assumption that, along with the term, came to dominate until the present day: that such an event could only have happened in the earth's distant past. By differentiating "abiogenesis," which included this qualifying proviso, from "spontaneous generation," Huxley finally succeeded in getting all of the amateur science and radical political implications of that earlier doctrine from around the neck of "Darwinism" as we have since come to know it. Furthermore, the importance of differentiating Brownian movement from true living movement became, in Huxley's hands, a weapon for the defeat of Bastian and "archebiosis," despite the fact that the very distinction at issue was first used by Bastian in support of archebiosis. Huxley's success cut off from the history of spontaneous generation disputes a previously significant related discourse: that of Brown's "active molecules," later called "histological molecules," observations of which were widely believed to support the possibility of heterogenesis.

As noted at the start, I do not intend to suggest that the experiments themselves were unimportant in the debate. The "dueling experiments" narrative of spontaneous generation debates has been well told, best of all by John Farley. Those experiments alone, however, were not sufficiently persuasive to determine the final marginalization of Bastian. The social context of the Darwinian scientific naturalists is also a crucial part of the story, and the spontaneous generation debates of the 1870s in Britain are misunderstood if they are seen only as a debate over experiments, or as a struggle between Darwinian science and outsiders.

Epilogue, 1880 through 1915

I have alluded to many continuities between discussion of spontaneous generation in the 1870s and discussion of the origin of life since the beginning of the twentieth century. An entire book would be needed to fill in the story from 1880 until the mid-twentieth century. Nonetheless, a few remarks are in order here until such a book appears.

Much previous work has emphasized a sharp break between the demise of spontaneous generation support in the 1870s and the resumption of the origin of life investigation by Alexander Oparin in the Soviet Union in 1924.[13] John Farley's book on the spontaneous generation controversies began to deconstruct this highly constructed discontinuity. I will argue further that Bastian's re-entry into publishing on the origin of life between 1900 and his death in 1915 was more influential as a source of continuity than even Farley's account recognizes.[14]

Two of Bastian's old adversaries in the debates of the 1870s, E. Ray Lankester and Edward Schäfer (later Sharpey-Shafer), subsequently became presidents of the British Association during the years in which Bastian revived his public advocacy of heterogenesis and archebiosis. Both of their presidential addresses, in 1906 and 1912 respectively, have been cited as summaries of the state of the origin of life question.[15] Schäfer's is much more frequently mentioned, since he devoted a considerably larger amount of his time to the origin of life.[16] Bastian's work during this period, on the other hand, is usually seen as a hopeless crusade, listened to by almost no scientist of any importance. He is often painted as a pitiful figure, unable to give up his private obsession despite the fact that the world paid no attention.[17]

It is true that the Royal Society refused to publish any of several papers Bastian submitted during these years[18] and that Bastian felt this scientific slight very deeply.[19] It would also not be accurate to make more than is justified of the notice that was taken. However, Bastian's publications of 1900–1915 are treated as though nobody responded to them at all. And that is a major source of the deceptive impression of discontinuity between 1880 and 1920, the time when Oparin commenced theorizing on the origin of life, that has dominated histories of the subject.

Despite the hegemony after 1880 of Huxley's terminology of "abiogenesis" and its built-in assumptions,[20] it was precisely because of Bastian's mechanistic, evolutionary approach that some twentieth-century researchers still saw Bastian as important. Even Bastian's old adversaries, Lankester and Schäfer, felt his challenge important enough to require a public rebuttal; Bastian was still an imposing figure that could not be ignored entirely. Schäfer addressed Bastian explicitly in his lecture, and Lankester remarked upon spontaneous generation only because Bastian had revived the issue.[21] In particular, Bastian revived his earlier emphasis on explaining the transition from nonlife to living in terms of colloid chemistry—a trend that grew markedly during these years.[22] This is not to argue that Bastian was the only source of this idea during the first decade of the twentieth century. The point I want to make is that, although picked up by several researchers at this time, including Bastian, Jacques Loeb, The Svedberg, and many others, the idea was not a new one but a revival and extension of an argument from the spontaneous generation debates of 1860–1880.[23] Some have insisted that this research program led up a blind alley and constituted a "Dark Age of Biocolloidollogy."[24] But one only dismisses such a huge body of work at the risk of being fundamentally ahistorical. Because this school of thought so dominated early twentieth-century thinking about the origin of life (through the late 1930s), and because it produced much data necessary to lead to more modern concepts, it is by definition a historical transition worthy of study.[25]

Furthermore, it is always hazardous to throw away a large body of carefully made observations, lest, once one comes to the correct interpretation, valuable data may be found therein. Microbiologist Milton Wainwright, for example, believes that much of what Bastian saw of microbial growth occurring in inorganic silica solutions can be interpreted as having been an early discovery of chemoautotrophy using silicon compounds.[26] Yet the microbiology of silicon was not understood until decades later—a delay could have been shorter if Bastian's experiments had not been so completely dismissed.

Among the scientists interested in the origin of life and influenced by Bastian between 1900 and 1915, several went on to become significant contributors to the field, including Ben Moore, Albert Mary, and Alfonso Herrera. Herrera provides an interesting example of continuity of ideas from the late nineteenth century into the mid-twentieth. He was an early and vociferous advocate of Darwinism in Mexico and, beginning in 1899, also began to investigate the origin of life. He carried out many of what came to be called "cell model experiments."[27] Herrera's work on the origin of life slipped into obscurity, along with the trend of research on colloid chemistry, by the time of his death in 1942. Oparin further minimized its importance in the 1950s.[28] However, since the 1960s several prominent researchers, especially Sidney Fox, have revived Herrera's reputation, crediting him with noting early the importance of such simple compounds as thiocyanide.[29]

Likewise, Felix D'Herelle's "micellar theory of life," derived from his study of bacteriophage, can be viewed as a direct descendant of speculations from 1860 to 1880 about colloidal particles as the ultimate units of life. Indeed, the term "micelles" that D'Herelle used for these ultimate colloidal particles was coined by Karl von Nägeli in his attempt to link up spontaneous generation with a theory of the structure of biological molecules.[30] Yet, because D'Herelle's theory was shown to be much less useful than the molecular biological explanation of bacteriophage pioneered by Luria and Delbrück, until very recently the earlier theory was relegated to the dustbin of history. Ignoring or minimizing these transitional ideas that predated an understanding of macromolecules led to an initial historiography of molecular biology that made that science seem to appear *de novo* out of thin air. At the very least, believers in evolutionary processes should regard such tales with skepticism.

In the same vein, the triumph of Cohn and Koch's monomorphist views led to complete dismissal of Nägeli and the pleomorphists (see chapters 5 and 6). In the 1870s, Koch *believed* monomorphism must be true.[31] Fortunately, Koch's assumption proved true to a first approximation, sufficient to launch his successful hunt for the pathogens of tuberculosis, wound infections, cholera, and many other major human killers.

But while the pleomorphists were wrong that bacterial species are totally illusory, Koch was also wrong in believing that stable species were incompatible with very extensive genetic mutability. Nonetheless, Koch's successes led to towering influence over the field and imposed monomorphist blinders on researchers in a way that delayed for several decades any investigation of just how great the limits of variation in bacteria are.[32] Observations of variable forms during those years were banished to such epistemological wastebasket categories as "involution forms," not seen as suitable for further study. This includes such phenomena as the smooth and rough variant forms of pneumococci, and, not so trivially, the resultant path to the double helix. By Koch's death in 1910 a few prominent bacteriologists had begun to publish observations of significant bacterial variability, but for many reasons the phenomenon was not recognized as significant by the mainstream until about 1920. Then research on variability enjoyed extraordinary prominence through the 1930s and much of the 1940s before receding into relative obscurity again.[33]

Many similar examples of dogmas have had their day: the "central dogma" of molecular biology, the "beads on a string" classical gene concept, the debates over which structures count as artifact (e.g., mitochondria, and golgi apparatus for many years) in staining cells, and later a whole new wave of such debates over artifact versus "real organelle" that came with the introduction of biological electron microscopy and its many preparatory procedures for the specimen.[34]

Lynn Margulis has told a similar story about the wealth of observations ignored in microscopic cytology in the early twentieth century that later led her to develop the cell symbiosis theory.[35] The constant presence of dogmatic ideas is a feature of science that should give us pause.[36] How many other similar stories are still waiting to be told?[37]

GLOSSARY

ACTIVE MOLECULE: Robert Brown's name (1828) for the bacteria-sized particles he saw produced from the breakdown in water of both living and nonliving matter.

BACTERIUM TERMO: Ehrenberg's name (1832) for the smallest bacterium visible with his microscope.

BIOPLASM: Beale's new name, after 1869 or so, for what he previously called "germinal matter" and to which he attributed the power of contagion.

CONFERVAE: green, strandlike algae. Strands from about the same size as *Leptothrix* (c. 1 μm in diameter), on up to thicker (eukaryotic) types such as *Spirogyra.*

GEMMULES: Darwin's minute "atoms of life" responsible in his hypothesis of pangenesis for conveying hereditary information into the gamete cells. The gemmules combined to form the gametes in a way very reminiscent of Buffon's "organic molecules" of 100 years earlier.

GERMINAL MATTER: Beale's name, from about 1864 to 1869, for tiny particles of living protoplasm that could multiply themselves. He considered those produced from the degeneration of diseased tissues to be the active agents of contagion.

GERM: everything from a resting or resistant spore to much more generally anything that could act as a structure analogous to a seed, only for microorganisms. To Tyndall, it was a particle so tiny as to be beyond any possibility of detection by microscopes.

LEPTOTHRIX: long chains of bacilli, strung together end to end.

MICROZYME: Burdon Sanderson's term, from about 1869 to 1874, for bacteria.

MOLECULE: term first used by Buffon to describe microscopic organic particles, in the size range of bacteria (c. 1–2 μm), that were said to be able to cluster together to form new cells. Used similarly by Brown (1828–1829), Schwann (1839), Addison (1841–1844), Bennett (1840s–1875), and many others, especially histologists and pathologists, and even by many opposed to spontaneous generation.

MONAD: a term used through the 1870s for single-celled organisms the size of protozoa (c. 10 to 500 μm). The "molecules" or "granules" within a monad, by contrast, were in the size range of bacteria, from 10 to 500 times smaller.

PENICILLIUM: mold fungi generally, i.e. those producing a mycelium, in some usages. In others, members of that particular genus of mold *(Penicillium)* only.

PLEOMORPHISM: the doctrine that bacteria do not come in distinct unchanging species, but because of their extreme mutability under varying environmental conditions can adopt very different forms. Views varied from belief that those changes of form were confined only within one or a few types of bacteria (and therefore did not undermine the entire project of constructing taxonomies of bacterial species) at one extreme, to the view of Nägeli or Bastian at the other extreme, that the mutability was so great that bacteria, yeasts, and fungi were completely intertransformable (and therefore no taxonomy of species made any sense). Huxley in 1870 believed in that more extreme view, but by 1875 or so, he no longer expressed that belief.

TORULA: the yeast form of fungi, e.g., in *Saccharomyces cerevisiae.*

VIBRIO: a bacterium with a slightly curled rodlike shape, similar to a sausage in form. Some used it almost interchangeably with "bacterium."

VIRUS: poison, generally.

ZOOGLEA: a cluster of cells embedded in a gelatinous matrix.

ZYMOSIS: the production of a communicable disease within the body of an organism infected by some material contagion. Most considered that the contagion had to be particulate or droplet in nature, as Graham's work on gas diffusion suggested that gases would disperse rapidly from their point of production and thus not remain concentrated enough to lead to epidemics.

ZYMADS: William Farr's (c. 1865–1867) term for the particles of contagion. Suggested they were chemical (not living) in nature.

TIMELINE

1859	November	Publication of Darwin's *On the Origin of Species.*
1860	April	English translation of Pouchet's "Atmospheric Micrography" appears.
		Owen's *Palaeontology* supports Pouchet.
	July	English translation of Pasteur's "New Experiments Relative to So-called Spontaneous Generation."
1861	January	English translation of Pasteur's following article, "New Expts. Relating to . . . Spontaneous Generation."
		Robert Grant's text *Recent Zoology* supports spontaneous generation and transformism.
		Pasteur awarded prize by French Academy for settling the question of spontaneous generation permanently.
1862	January	Darwin writes favorably of Pasteur's new *Memoir sur les Corpuscles* in letter to Henry Holland.
	May	Jeffries Wyman's experiments appearing to support the possibility of spontaneous generation published in *American Journal of Science (Silliman's Journal).*
		Huxley's *Lectures to Working Men* declares Pasteur's work conclusive.
1863	October	Bennett, after reading Pasteur, begins a series of experiments on spontaneous generation in Edinburgh; Child begins experiments in Oxford at about the same time.
1864	June	Child's first paper read to Royal Society, appears to support possibility of spontaneous generation. In July, his influential review is published in *British and Foreign Medico-Chirurgical Review.*
	September	Schaaffhausen's experiments in Bonn reported in *British Medical Journal*; X Club formed.
	December	*Lancet* supports heterogenesis.
1865	November	Burdon Sanderson and Beale begin experiments on the origin of Cattle Plague.
1867		Wyman's second series of experiments published, suggesting that experimental error caused the appearance of spontaneous generation in his first series of 1862 and that boiling for five hours prevented it in all flasks.

		Lister announces his system of antiseptic surgery in *Lancet*.
1868	January	Darwin's *On the Variation of Animals & Plants* appears, announcing pangenesis hypothesis.
	March	Bennett's "Lecture on the Atmospheric Germ Theory" published in *Edinburgh Medical Journal*.
	August	Huxley announces discovery of *Bathybius* at BAAS meeting.
	November	Huxley gives a lecture, "The Physical Basis of Life," which is published in February 1869; Huxley begins experiments on bacteria, yeast, and mold and writes to warn Haeckel against materialism.
		Owen's *On the Anatomy of Vertebrates*, v. 3 appears.
	December	Spencer writes to deny that evolution or his physiological units have anything to do with spontaneous generation.
1869	January	*British Medical Journal* series "On the Origin of Life" begins, continuing through December.
	May	Beale attacks Huxley on "Protoplasm."
1870	January	Tyndall's "Dust and Disease" lecture, reviewed harshly by *Lancet* and *British Medical Journal*.
	April	Tyndall and Bastian debate via letters in the *Times*.
	May	Huxley seems to conclude that Bastian and his work are untrustworthy.
	June/July	Bastian's first articles appear in *Nature*.
	September	Huxley's Liverpool BAAS address "Biogenesis and Abiogenesis." Huxley attacks Bastian.
1871	April	Bastian's book *Modes of Origin of Lowest Organisms* responds to Huxley.
	August	Thomson's BAAS address on the meteoric origin of life.
1872	June	Bastian's *Beginnings of Life* published.
	August	Wallace reviews *Beginnings of Life* very favorably in *Nature*.
	December	Burdon Sanderson and Bastian conduct experiments together.
1873	January	Burdon Sanderson supports Bastian's capabilities as an experimenter in *Nature*; Lankester and Roberts respond critically.
	June	Pode and Lankester publish their criticism in a paper to Royal Society; Murchison supports Bastian in a new edition of *Continued Fevers of Great Britain*.
	August	First installment of Dallinger and Drysdale experiments, "Life History of the Monads," appears in *Monthly Microscopical Journal*.

1874	April	Roberts paper read to Royal Society opposing Bastian, but holding out possibility of abiogenesis in rare circumstances; publication of Bastian's *Evolution and the Origin of Life*.
		Death of Robert Grant.
1875	April	Debate in Pathological Society supports Bastian and casts doubt on germ theory of disease.
	June	Huxley renounces *Bathybius* as artifact.
	September	Death of Bennett.
	October	Tyndall begins new series of experiments to disprove Bastian, through January 1876.
1876	January	Tyndall reports totally negative results to Royal Society on spontaneous generation in infusions, "though given 500 chances."
	May	Bastian's paper in reply to Tyndall read at Royal Society (never published).
	July	Cohn's *Beiträge* article announcing discovery of heat-resistant spores in hay bacillus.
	late September	Cohn visits London, hands Tyndall the article on heat-resistant spores and Koch's anthrax paper.
	6 October	Tyndall begins new experiments on infusions, finding that a great many cannot be sterilized.
	19 October	In talk in Glasgow, Tyndall describes Koch's work in detail (before Pasteur has heard of Koch).
	October–early January 1877	Tyndall cannot succeed in sterilizing infusions at the Royal Institution.
1877	early January	Tyndall moves experiments to Kew Gardens; succeeds in getting sterile infusions.
	February	Tyndall announces that he can get sterile infusions at Royal Institution again, but only using repeated short boilings (fractional sterilization).
	May	Tyndall's major technical paper published in *Philosophical Transactions of the Royal Society of London*. Most, including Darwin, conclude that heat-resistant spores provide the final explanation for why Bastian must be wrong.
	June	Royal Society will not allow Bastian's paper in reply, so Bastian reads it to Linnean Society.
1878	Spring	Final exchange of articles between Tyndall and Bastian, in the *Nineteenth Century*. Bastian becomes full professor at University College Medical School.
1880		Publication of Bastian's *The Brain as an Organ of Mind*; the book still supports heterogenesis and archebiosis.
1881		Bastian and Pasteur disagree at International Medical Congress in London.

CAST OF CHARACTERS

BASTIAN, Henry Charlton (1837–1915). Professor of Pathological Anatomy at University College London (UCL) Medical School, 1867–1887. Afterwards Professor of Medicine at UCL from 1887 to 1898. He was also physician to the National Hospital, Queen Square, from 1868 to 1902 and full physician to University College Hospital from 1878 to 1898. Bastian was the most ardent, vocal, and prolific British experimenter and writer in support of spontaneous generation, from 1869 to 1915. After initial experiments confirmed his view that evolutionary theory required the possibility of a transition from inanimate to living matter, Bastian published those experiments in *Nature* in June-July 1870. After Huxley attacked his views in September 1870, he responded in an 1871 book, *The Modes of Origin of Lowest Organisms,* and a much larger, two-volume 1872 work, *The Beginnings of Life.* Bastian's reputation as a competent experimenter was attacked first by Huxley (1870–1873), then by Tyndall (1875–1878), causing Bastian to withdraw from controversy on the issue until his retirement from UCL in 1898. From then until his death he published more voluminous experimental works on heterogenesis and archebiosis.

BEALE, Lionel Smith (1828–1906). Professor at Kings College and one of Britain's most renowned experts on microscopy. Beale was a staunch vitalist and thus opposed all spontaneous generation, though his own theory of "bioplasm" seemed tantamount to heterogenesis to many. Beale's irascible nature led him to make enemies of many who might have been his allies in opposing spontaneous generation. He and Tyndall had greater mutual enmity, seemingly, than either did for Bastian.

BENNETT, John Hughes (1812–1875). Professor of the Institutes of Medicine (physiology) at Edinburgh, 1848–1874. Bennett's "molecular theory" of cell structure and cell formation was widely known by the early 1860s. In 1863, after a heated critique of this theory by Beale, Bennett seems to have realized what he had previously denied: that the theory was tantamount to heterogenesis. He began to repeat some of Pasteur's and Pouchet's experiments, and by 1868 declared himself convinced that the experiments supported Pouchet. He continued to publish in this vein until 1872, shortly before he died.

BUSK, George (1807–1886). Originally trained in medicine, Busk became a zoologist and teacher of biology in 1854. President of the Linnean Society when Bastian first became a member in the early 1860s, he gave Bastian support both there and at the Royal Society. A member of the X Club.

DALLINGER, William Henry (1842–1909). Methodist minister and biologist with much talent in microscopy. Dallinger opposed spontaneous generation and carried out some influential experiments (with Drysdale) suggesting that monads (protists) had life cycles with stages of differing appearances, including very tiny spores sometimes capable of remarkable heat resistance. These experiments, conducted between 1873 and 1875, were popularized by Tyndall, Darwin, and Lawson in 1876 and 1877 in opposition to Bastian.

DARWIN, Charles (1809–1882). Originally he toyed with Lamarckian ideas of spontaneous generation, especially under the tutelage of Robert Grant, but by 1838 Darwin no longer considered the idea central to his evolutionary theory. Nonetheless, he continued to take a lively interest in new experiments on the question: he considered Pasteur's experiments convincing in 1863, but then found Bastian's work persuasive in 1872. He does not seem to have accepted the concept of spontaneous generation as finally refuted until Tyndall's work of 1877.

DOHRN, Anton (1840–1909). Head of the Naples Marine Research Station and an ardent supporter of Darwin and Huxley. His Marine Station for experimental biology served as a model for many others around the world, including Woods Hole, Massachusetts. Dohrn from the first opposed Bastian and spontaneous generation. He was a close friend of E. Ray Lankester.

DRYSDALE, John (1817–1892). Liverpool physician who carried out important microscopic work on life cycles in monads (protists) along with William Dallinger, 1873–1875. Published *On the Protoplasmic Theory of Life* in 1874.

FISKE, John (1842–1901). Professor of Philosophy at Harvard University. Fiske was an enthusiast of Darwinism and especially of Herbert Spencer's philosophy. In 1873, he wrote that Bastian and Huxley were equally devoted Darwinians and equally scientific in their opposing positions on spontaneous generation. By late 1875 he had taken a more hardline position against Bastian, in accordance with that of Huxley and Tyndall.

FRANKLAND, Edward (1825–1899). Professor of Chemistry at the Royal Institution, London, 1863–1869; Professor of Chemistry at Royal College of Chemistry, 1865–1885. X Club member who did much research on water chemistry and microbiology. He was an early participant in Bastian's experiments through much of 1870.

HOOKER, Sir Joseph Dalton (1817–1911). Botanist, Director of the Royal Botanic Gardens at Kew. Member of the X Club and longtime friend of Huxley, he opposed spontaneous generation but was an enthusiast of Huxley's pleomorphist ideas of 1870.

HUXLEY, Thomas Henry (1825–1895). Professor of Natural History in the Government School of Mines and a naturalist with the Geological Survey. Also Secretary of the Royal Society from 1871 to 1880 and President of the Royal Society from 1883 to 1885. He was thought by much of the British public to oppose any kind of religious opposition to evolution and a naturalistic worldview; thus many believed he privately supported spontaneous generation, especially after his discovery of *Bathybius* and his lecture in 1868, "The Physical Basis of Life." He publicly opposed Bastian and spontaneous generation beginning in September 1870 in his address to the BAAS, "Biogenesis and Abiogenesis." From that point on, he insisted that evolution did logically call for abiogenesis, but only in the earth's distant past.

LANKESTER, E. Ray (1847–1929). Became editor, following his father, of the *Quarterly Journal of Microscopical Science* in 1869. Became Professor of Zoology at University College London upon Grant's death in 1874. Lankester was an extremely talented embryologist and comparative anatomist, remaining at University College until 1891, when he became Professor of Comparative Anatomy at Oxford University. He also became Director of Natural History at the British Museum in 1898. He held these last two posts until 1907. He was from the first an enthusiastic supporter of abiogenesis in the distant past as an essential corollary of evolution. But he was an equally staunch, even vicious, opponent of Bastian and present-day spontaneous generation, still attacking Bastian as late as 1906 in his BAAS presidential address.

LAWSON, Henry (1841–1877). Lecturer in histology at St. Mary's Hospital Medical School and editor of *Monthly Microscopical Journal, Popular Science Review,* and *Scientific Opinion.* Lawson was an enthusiastic supporter of Darwin and of heterogenesis. He editorialized in favor of Pouchet, Bennett, Owen, and Bastian from 1868 to 1870. By October 1871 he was distancing himself from his earlier support for Bastian and heterogenesis, and by 1873 was strongly opposing Bastian and supporting Huxley.

LISTER, Joseph (Lord) (1827–1912). Regius Professor of Surgery at University of Glasgow, 1860–1866; Professor of Surgery at University College London, 1866–1869; Professor of Clinical Surgery at Edinburgh, 1869–1876; Professor of Clinical Surgery at King's College, London beginning in 1876. Lister first described antiseptic surgery in 1867 and believed Pasteur's disproof of spontaneous generation indicated germs as the cause of wound infections. He was a believer in pleomorphism of bacteria through most of the 1870s, but an opponent of spontaneous generation.

MACMILLAN, Alexander (1818–1896). Influential British publisher who supported the Darwinians and published the journal *Nature.* He was a major supporter of Bastian, even in the face of opposition from the X Club, publishing all of Bastian's works on spontaneous generation through the 1870s and contracting the American rights to those books to Edward Youmans.

OWEN, Richard (1804–1892). Britain's most celebrated comparative anatomist, in charge of the Hunterian collections at the Royal College of Surgeons. He was a long-time friend of Pouchet and declared support for heterogenesis in 1860, at first in a very quiet way. By 1868, now a bitter enemy of Huxley and the Darwinians, he made a major statement in favor of spontaneous generation in his *On the Anatomy of the Vertebrates*. Although he made no further public statements on the subject, he was widely cited as a major supporter of spontaneous generation throughout the early 1870s.

PASTEUR, Louis (1822–1895). Professor of Chemistry and Dean of Sciences at the University of Lille, France, 1854–1857; Assistant Director of Scientific Studies at École Normale Supérieure, Paris, 1857–1867; Professor of Chemistry at University of Paris (Sorbonne), 1867–1874; Director of Laboratory of Physiological Chemistry, École Normale, Paris, 1867–1888. Pasteur first performed prize-winning experiments opposing spontaneous generation in 1860–1862. He continued to oppose later supporters of the doctrine, including Bastian from 1876 to 1877, although he was unaware of the existence of heat-resistant spores and their role in the experiments until Tyndall brought this to his attention in late 1876.

POUCHET, Felix Archimede (1800–1872). French naturalist, corresponding member of the Academy of Sciences, and specialist on animal generation at the Rouen Natural History Museum. His 1859 book *Hétérogenie* advanced theoretical and experimental support for heterogenesis that led to a series of celebrated public debates between Pouchet and Louis Pasteur from 1860–1864. The French Academy declared Pasteur the winner, though many in France and Britain considered this unjust.

SANDERSON, John Burdon (1828–1905). First lecturer, then Professor of Physiology at University College London, 1870–1883, and at Oxford, 1883–1895. His reputation in British physiology was second only to Sharpey in the 1870s, and later to Michael Foster. He carried out with Bastian replications of some of the latter's key experiments, especially turnip-cheese infusions, from December 1872 to January 1873. While he steadfastly refused to say the experiments must prove archebiosis valid, he insisted that Bastian's experimental technique was beyond reproach and that after thorough boiling in sealed flasks, bacteria could be seen growing abundantly in many of Bastian's infusions.

SCHÄFER, Edward (1850–1935). Originally Sharpey and Burdon Sanderson's assistant in physiology at University College London, he became Assistant Professor of Physiology there in 1874 and full Professor from 1883 to 1899, then Professor of Physiology at Edinburgh, 1899–1933. He was a founding member of the Physiological Society in 1876 and an opponent of spontaneous generation from his college days, still criticizing Bastian's experiments as late as his BAAS presidential address in 1912. He later changed his last name to Sharpey-Schafer.

SHARPEY, William (1802–1880). Professor of Physiology and General Anatomy at University College London, 1836–1874. England's most influential physiology teacher and sometime Secretary for Biological Sciences to the Royal Society until 1871. He was never a supporter of spontaneous generation, but did respect Bastian and assist his career.

SPENCER, Herbert (1820–1903). X Club member and author of the multivolume

Synthetic Philosophy, including evolution. Spencer avowed in 1868 that he had been misunderstood by those who thought his "physiological units" were an idea verging upon spontaneous generation. Although from that moment he opposed spontaneous generation, he became a close and lifelong friend of Bastian, who was the executor of Spencer's will. Both men shared an interest in the origin of the elements by an evolutionary process, one of the earliest steps in Spencer's philosophical system.

THOMSON, William (Lord Kelvin) (1824–1907). Professor of Natural Philosophy (physics) at Glasgow University, 1846–1899. Thomson was one of the most renowned physicists of the nineteenth century and an ardent opponent of evolution and of spontaneous generation. His 1871 BAAS presidential address went so far as to suggest that life could have come to earth on meteorites, in order to sidestep giving a naturalistic explanation for the first origin of life.

TYNDALL, John (1820–1893). Professor of Natural Philosophy (physics) at the Royal Institution, London, 1853–1893. Superintendent of the Royal Institution after Faraday's death in 1867. Also Professor of Physics at the Government School of Mines. He was an early convert to the germ theory of disease, in 1869, and from that moment onwards became an obdurate opponent of any experimental work that ever seemed to suggest the possibility of spontaneous generation. He first proclaimed his position in a January 1870 lecture, "Dust and Disease." In April 1870 he and Bastian first disagreed over the issue publicly, in an exchange of letters in the London *Times.* He undertook extensive experimental work from 1875 to 1877 to try to disprove Bastian conclusively. This resulted in his discovery in January 1877 of fractional sterilization as a means to kill heat-resistant bacterial endospores.

WALLACE, Alfred Russel (1823–1913). Co-discoverer, with Darwin, of natural selection. He was an early supporter of Bastian's work, in 1872, and recommended it highly to Darwin. In later years, as Wallace discovered that his interest in spiritualism was incommensurable with a belief in a materialistic abiogenetic origin of life, Wallace repudiated Bastian's work in this area, though still holding him in high esteem as an evolutionary thinker.

YOUMANS, Edward L. (1821–1887). American publisher and supporter of evolutionary science. Youmans published the works of many of the X Club, as well as of Bastian, in American editions. He saw the debate among evolutionists over spontaneous generation as a healthy dispute over a genuinely open question, important to the coherence of Darwinian theory. Thus, he encouraged Bastian to keep goading Spencer, Huxley, and Tyndall on the issue.

NOTES

Abbreviations

ARWLR James Marchant, ed., *Alfred Russel Wallace: Life and Reminiscences* (New York: Harper & Bros., 1916), 2 vols.
BFMCR *British and Foreign Medico-Chirurgical Review*
BMJ *British Medical Journal*
CCD Frederick Burkhardt et al., eds., *The Correspondence of Charles Darwin*, vols. 10 and 11 (Cambridge: Cambridge University Press, 1997 and 2000).
DCC Frederick Burkhardt et al., eds., *A Calendar of the Correspondence of Charles Darwin, 1821–1882, with Supplement* (Cambridge: Cambridge University Press, 1994).
DNB *Dictionary of National Biography*
LLCD Francis Darwin, ed., *Life and Letters of Charles Darwin* (London: Macmillan, 1887), 3 vols.
LLCL K. M. Lyell, ed., *Life, Letters and Journals of Sir Charles Lyell, Bart.* (London: Macmillan, 1881), 2 vols.
LLEF Edward Frankland, *Sketches from the Life of Edward Frankland* (London: Spottiswoode, 1902).
LLGJR E. Romanes, ed., *Life and Letters of George John Romanes* (London: Longmans Green, 1896).
LLHS David Duncan, ed., *Life and Letters of Herbert Spencer* (New York: Appleton, 1908), 2 vols.
LLJDH Leonard Huxley, ed., *Life and Letters of Sir Joseph Dalton Hooker* (New York: Appleton, 1918), 2 vols.
LLTHH Leonard Huxley, ed., *Life and Letters of Thomas Henry Huxley* (London: Macmillan, 1900), 2 vols.
LLWT A. Logan Turner, *Sir William Turner: A Chapter in Medical History* (Edinburgh: Wm. Blackwood and Sons, 1919).
LWJT A. S. Eve and C. H. Creasey, *Life and Work of John Tyndall* (London: Macmillan, 1945).
MLCD Francis Darwin and A. C. Seward, eds., *More Letters of Charles Darwin* (London: Macmillan, 1903), 2 vols.
MMJ *Monthly Microscopical Journal*

PRSL	*Proceedings of the Royal Society of London*
PTRSL	*Philosophical Transactions of the Royal Society of London*
QJMS	*Quarterly Journal of Microscopical Science*

Introduction

1. John Farley's masterful analysis shows clearly the extent to which the outcome of these disputes was almost always underdetermined by the experimental evidence alone; see Farley, *The Spontaneous Generation Controversy from Descartes to Oparin* (Baltimore: Johns Hopkins University Press, 1977), especially chapters 5 and 7. On how the resolution of these debates was related to the birth of microbiology, see several detailed internalist histories: William Bulloch, *History of Bacteriology* (London: Oxford University Press, 1938); E. F. Gale, "The Development of Microbiology," in Joseph Needham, ed., *The Chemistry of Life* (Cambridge: Cambridge University Press, 1970), pp. 38–59; also G. Baldacci, L. Frontali, and A. Lattanzi, "The Debate on Spontaneous Generation and the Birth of Microbiology," *Fundamenta Scientiae* 2: 123–136 (1981).
2. Harold Abrahams, "Priestly Answers the Proponents of Abiogenesis," *Ambix* 12: 44–71 (1964), p. 45.
3. Paula Findlen, "Controlling the Experiment: Rhetoric, Court Patronage and the Experimental Method of Francesco Redi," *History of Science* 31: 35–64 (1993). For more detail on the period up to and including Redi's work, see Everett Mendelsohn, "Philosophical Biology versus Experimental Biology: Spontaneous Generation in the Seventeenth Century," in Marjorie Grene and Everett Mendelsohn, eds., *Topics in the Philosophy of Biology, Boston Studies in the Philosophy of Science* 27: 37–65 (1976); see also Catherine Wilson, *The Invisible World: Early Modern Philosophy and the Invention of the Microscope* (Princeton: Princeton University Press, 1995), esp. pp. 114–118 and 199–205.
4. John Farley, "The Political and Religious Background to the Work of Louis Pasteur," *Annual Reviews of Microbiology* 32: 143–154 (1978). It is worth noting that Nils Roll-Hansen has argued against the externalist analysis of Farley and of Geison. See "Experimental Method and Spontaneous Generation: The Controversy between Pasteur and Pouchet, 1859–64," *Journal of the History of Medicine and Allied Sciences* 34: 273–292 (1979); also "Pasteur: An Underestimated Hero of Science," *Centaurus* 40: 81–93 (1998). See also Antonio Gálvez, "The Role of the French Academy of Sciences in the Clarification of the Issue of Spontaneous Generation in the mid-Nineteenth Century," *Annals of Science* 45: 345–365 (1988). Gerald Geison has responded to these critics in his book *The Private Science of Louis Pasteur*

(Princeton: Princeton University Press, 1995), p. 321. This response exactly sums up my own sense of the debate. It is not about the facts of what happened; rather, it is about whether one fundamentally believes that factors beyond the data actually participate in the making of scientific knowledge. An intermediate position has been taken by Iris Fry in *The Emergence of Life on Earth: A Historical and Scientific Overview* (New Brunswick, N.J.: Rutgers University Press, 2000), pp. 46–53.

5. T. H. Huxley (1870), "Biogenesis and Abiogenesis," pp. 232–274 in *Discourses Biological and Geological, Essays* (New York: Appleton, 1925), p. 239.
6. Shirley Roe, "Buffon and Needham: Diverging Views on Life and Matter," pp. 439–450 in *Buffon '88* (Paris and Lyon: J. Vrin, 1992). See also chapter 5 of Clara Pinto-Correia, *The Ovary of Eve* (Chicago: University of Chicago Press, 1997); and Marguerite Carozzi, "Bonnet, Spallanzani and Voltaire on Regeneration of Heads in Snails: A Continuation of the Spontaneous Generation Debate," *Gesnerus* 42: 265–288 (1985).
7. Shirley Roe, "John Turberville Needham and the Generation of Living Organisms," *Isis* 74: 159–184 (1983).
8. Philip Sloan, "Organic Molecules Revisited," pp. 415–438 in *Buffon '88*, (Paris and Lyon: J. Vrin, 1992).
9. Ibid.
10. J. B. Lamarck, *Zoological Philosophy* (Chicago: University of Chicago Press, 1984), esp. pp. 236–248.
11. Richard Burkhardt, *The Spirit of System: Lamarck and Evolutionary Biology*, (Cambridge, Mass.: Harvard University Press, 1977), especially chapters 4 and 5.
12. The same associations were made all over Europe. See Sander Gliboff, "Evolution, Revolution and Reform in Vienna: Franz Unger's Ideas on Descent and Their Post-1848 Reception," *Journal of the History of Biology* 31: 179–209 (1998).
13. Adrian Desmond, *Politics of Evolution* (Chicago: University of Chicago Press, 1989); Desmond and James R. Moore, *Darwin* (London: Michael Joseph, 1991).
14. See Alison E. Adam, "Spontaneous Generation in the 1870s: Victorian Scientific Naturalism and its Relation to Medicine," Ph.D. diss., Sheffield Hallam University, 1988, pp. 51–53.
15. Many writers at the time lamented this situation, for example James Samuelson, "On the Source of Living Organisms," *Quarterly Journal of Science* 1: 598–614 (1864), pp. 606, 614.
16. John Lowe, "On *Sarcina ventriculi* Goodsir," *Edinburgh New Philosophical Journal* 12: 58–64 (n.s., 1860), pp. 61–62 gives a good early example antic-

ipating the later, more developed pleomorphist theories of Huxley, Lankester, and Lister.

17. There were many different "germ theories" of disease, for example, which I discuss in chapter 6.
18. Ludwik Fleck, *The Genesis and Development of a Scientific Fact,* English trans. by Fred Bradley and Thaddeus Trenn (Chicago: University of Chicago Press, 1979), p. 43.
19. Adam, "Spontaneous Generation in the 1870s." On Robert Chambers's popularization of science in *Vestiges* as opposed to Huxley's strategy of public science education, see Joel Schwartz, "Robert Chambers and Thomas Henry Huxley, Science Correspondents: The Popularization and Dissemination of Nineteenth Century Natural Science," *Journal of the History of Biology* 32: 343–383 (1999).
20. See Glenn Vandervliet, *Microbiology and the Spontaneous Generation Debate During the 1870s* (Lawrence, Kans.: Coronado Press, 1971); John Crellin, "The Problem of Heat-Resistance of Microorganisms in the British Spontaneous Generation Controversies of 1860–1880," *Medical History* 10: 50–59 (1966).
21. The attempt is for as fully detailed a portrait as possible, given the lack of preservation of Bastian's personal papers. After two years of searching for surviving heirs, I ascertained that those papers were lost or sold to collectors over the years. The number of Bastian letters I have located in various extant collections (Huxley, Lockyer, Macmillan, Wyman, Wellcome, University College, Royal Society, Linnean Society, etc.) probably represents the most complete sample assembled to date.

1 Spontaneous Generation and Victorian Science

1. This argument was made by Spencer in response to some who interpreted his writing on "physiological units" to be a plea for spontaneous generation as part of evolutionary doctrine. See his Dec. 1868 essay, "On Alleged 'Spontaneous Generation,' and on the Hypothesis of Physiological Units," appended to the 2nd ed. of his *Principles of Biology* (New York: Appleton, 1870), pp. 479–492. Darwin himself did not think his theory stood or fell based on a clear answer to the question of how life originated. See John Farley, *The Spontaneous Generation Controversy from Descartes to Oparin* (Baltimore: Johns Hopkins University Press, 1977), p. 81 and Darwin to Lyell, 18 Feb. 1860 in *MLCD* 1: 140–141.
2. Pouchet was a believer in heterogenesis, but not in the more radical doctrine of abiogenesis or archebiosis. Henry Charlton Bastian believed in both; Huxley and John Tyndall in neither, at least not during present times.

3. See Robert Grant, *Recent Zoology* (London: Walton and Maberly, 1861), pp. 5–6, 9.
4. Quoted in *LLCD,* v. 2, p. 188.
5. Thomas H. Huxley, "Biogenesis and Abiogenesis" (1870), pp. 572–594 in *The Scientific Memoirs of Thomas Henry Huxley,* v. 3 (London: Macmillan, 1901); Huxley's paper can also be found in his *Discourses Biological and Geological, Essays* (New York: Appleton, 1925), pp. 232–274 (pagination from this edition used hereafter).
6. Huxley, "On Some Organisms Living at Great Depths in the North Atlantic Ocean," *QJMS* 8: 203–212 (1868).
7. Huxley, "The Physical Basis of Life," lecture to working men in Edinburgh, Nov. 1868, published Feb. 1869 in *Fortnightly Review* 11 (5 n.s.): 129–145. One of the reasons Huxley's talk became so controversial was its publication in a review that reached a broad educated audience. The controversy caused interest to snowball, so that this edition of the *Fortnightly* ran to extra printings and a total of five or six times as many copies as any other issue in that decade.
8. Joseph Lister, "On the Antiseptic Principle of the Practice of Surgery," *Lancet* 1: 326–329, 357, 387, 507; 2: 95 (1867).
9. John Tyndall, "Dust and Disease," Jan. 1870 lecture at the Royal Institution. Originally published as "On Haze and Dust" in *Nature* 1: 339–342 (27 Jan. 1870).
10. John Burdon Sanderson, "Dr. Bastian's Experiments on the Beginnings of Life," *Nature* 7: 180–181 (1873), p. 180.
11. Darwin to John Tyndall, 4 Feb. 1876, Tyndall papers, tss. p. 2850; *DCC* #10379.
12. Bentham, "Mr. Bentham's Anniversary Address to the Linnean Society," *Nature* 6: 131–133 (1872), p. 132.
13. Bastian, "Spontaneous Generation," *Nature* 9: 482–483 (1874).
14. E. Ray Lankester, "Dr. Sanderson's Experiments," *Nature* 7: 242–243 (1873); also C. C. Pode and Lankester, "Experiments on the Development of *Bacteria* in Organic Infusions," *PRSL* 21: 349–358 (1873).
15. William Roberts, "Dr. Bastian's Experiments on the Beginning of Life," *Nature* 7: 302 (1873).
16. Roberts, "Studies on Abiogenesis," *PTRSL* 164: 457–477 (1874).
17. Ibid., p. 462.
18. Ibid., p. 463.
19. Ibid., p. 477.
20. Justus von Liebig, *Organic Chemistry in its Applications to Agriculture and Physiology* (London: Taylor and Walton, 1840).
21. John Burdon Sanderson, "Introductory Report on the Intimate Pathology

of Contagion," Appendix to *Twelfth Report of the Medical Officer of the Privy Council, 1869* (London, 1870), pp. 229–256. Burdon Sanderson's first name for these objects invokes the work of Antoine Béchamp on "microzymas," which began in the 1860s and was reported in the British scientific press at least as early as 1868.

22. Tyndall, "Professor Tyndall on Filtered Air," *Times,* 7 Apr. 1870.
23. Bastian, "The Germ Theory of Disease," *Times,* 13 Apr. 1870.
24. Tyndall, "The Germ Theory of Disease," *Times,* 21 Apr. 1870, p. 8.
25. Bastian, "To the Editor of the Times," *Times,* 22 Apr. 1870, p. 5.
26. Ferdinand Cohn, "Untersuchungen über Bacterien. IV.," *Beiträge zur Biologie der Pflanzen* 2(2): 249–276 (1876).
27. Tyndall, "Preliminary Note on the Development of Organisms in Organic Infusions," *PRSL* 25: 503–506 (1877).
28. Tyndall, "On Heat as a Germicide when Discontinuously Applied," *PRSL* 25: 569 (1877). The technique has since come to be called "tyndallization."
29. See Bernard Becker, *Scientific London* (New York: Appleton, 1875) on the more established London scientific societies in this period.
30. See Jan Golinski, *Science as Public Culture* (Cambridge: Cambridge University Press, 1992), for a discussion of the early history of the Royal Institution and science demonstrations for the public.
31. On the Linnean Society, see Andrew Gage and William Stearn, *A Bicentennary History of the Linnean Society of London* (London: Academic Press, 1988).
32. On the early years of the BMA, see Peter Bartrip, *Themselves Writ Large: The British Medical Association, 1832–1966* (London: BMJ Publishing Group, 1996).
33. On University College London and its founding, see H. Hale Bellot, *University College London* (London: London University Press, 1929); also David Mabberley, *Jupiter Botanicus: Robert Brown of the British Museum* (Braunschweig: Verlag von J. Cramer, 1985), pp. 262–268. On Grant's radical politics and medical reform agenda see Adrian Desmond, *Politics of Evolution* (Chicago: University of Chicago Press, 1989). On Wakley and the *Lancet* and medicine during this period in Britain, see ibid.; also W. L. Burn, *The Age of Equipoise: a Study of the Mid-Victorian Generation* (New York: Norton, 1965), pp. 202–216.
34. Adrian Desmond and James R. Moore, *Darwin* (London: Michael Joseph, 1991).
35. William Paley's *Natural Theology; or Evidences of the Existence and Attributes of the Deity Collected from the Appearances of Nature* had been the dom-

inant work among British intellectuals on this subject for over a generation.

36. On these debates, see Toby Appel, *The Cuvier-Geoffroy Debate* (Oxford: Oxford University Press, 1987).
37. See Desmond, *Politics* on Owen's conflicts with Grant; Evelleen Richards, "A Question of Property Rights," *British Journal for the History of Science* 20: 129–171 (1987) on the development of Owen's evolutionary thinking; also Nicolaas Rupke, *Richard Owen: Victorian Naturalist* (New Haven: Yale University Press, 1994) on Owen's entire career.
38. Robert E. Grant, *Recent Zoology: a Tabular View of the Primary Divisions of the Animal Kingdom* (London: Walton and Maberly, 1861).
39. Ray Society Minute Book, meetings of 5 Nov. 1847, f. 127–128; 7 Dec. and 17 Dec. 1847, ff. 132–137. Also Thomas Bell to William Jardine, Minute Book ff. 130–131, BMNH (Zoology). Richards has discussed this episode in "A Question of Property Rights," pp. 163–165, see also pp. 155–156.
40. Rupke, *Richard Owen*, pp. 97–105. Contrast this with Adrian Desmond's portrayal in *Huxley: From Devil's Disciple to Evolution's High Priest* (Reading, Mass.: Addison-Wesley, 1997).
41. See Roy MacLeod, "The X Club," *Notes and Records of the Royal Society* 24: 305–322 (1970); and Ruth Barton, "'An Influential Set of Chaps': The X-Club and Royal Society Politics 1864–85," *British Journal for the History of Science* 23: 53–81 (1990); also Barton, "Huxley, Lubbock and a Half-Dozen Others," *Isis* 89: 410–444 (1998) and Colin Russell, *Edward Frankland: Chemistry, Controversy and Conspiracy in Victorian England* (Cambridge: Cambridge University Press, 1996), chapter 11.
42. A good account of Tyndall and his activity to help define science and secure its cultural authority can be found in Thomas Gieryn, *Cultural Boundaries of Science* (Chicago: University of Chicago Press, 1999), chapter 1. See also Frank Turner, *Contesting Cultural Authority* (Cambridge: Cambridge University Press, 1993), p. 204.

2 "Molecular" Theories

1. Robert Brown, *A Brief Account of Microscopical Observations made on the Particles Contained in the Pollen of Plants, and on the General Existence of Active Molecules in Organic and Inorganic Bodies*, pamphlet (found in Heidelberg University Library) donated by Brown to the Royal Academy of Sciences, Munich. Also published in *Philosophical Magazine* 4: 161–173 (1828). See also Brown, "Additional Remarks on Active Molecules," *Philosophical Magazine* 6: 161–166 (1829).

2. Henry Bence Jones, *Life and Letters of Faraday* (London: Longmans Green, 1870), v. 1, pp. 403–404; John B. Dancer, "Remarks on Molecular Activity as Shown Under the Microscope," *Proceedings of the Literary and Philosophical Society of Manchester* 7: 162–164 (1868); Steven Brush, "Brownian Movement from Brown to Perrin," *Archives of History of the Exact Sciences* 5: 1–36 (1968); David Goodman, "The Discovery of Brownian Motion," *Episteme* 6: 12–29 (1972). Goodman shows that the belief that Brown had implied self-active molecules capable of spontaneous generation lasted for several years.
3. William Coleman, "Cell, Nucleus, and Inheritance: An Historical Study," *Proceedings of the American Philosophical Society* 109: 124–158 (1965); John Crellin, "The Dawn of the Germ Theory: Particles, Infection and Biology," pp. 57–76 in F. N. L. Poynter, ed., *Medicine and Science in the 1860s* (London: Wellcome Institute, 1968); Steven Brush, "Brownian Movement"; Russell Maulitz, "Schwann's Way: Cells and Crystals," *Journal of the History of Medicine and Allied Sciences* 26: 422–437 (1971); Leland Rather, *Addison and the White Corpuscles* (London: Wellcome Institute, 1972), pp. 218–220; Rather, *The Genesis of Cancer* (Baltimore: Johns Hopkins, 1979), pp. 67–68.
4. John Snow, *On Continuous Molecular Changes* (London: John Churchill, 1853), p. 145.
5. John Hughes Bennett, "On the Molecular Theory of Organisation," *Proceedings of the Royal Society of Edinburgh* 4: 436–446 (1861). This doctrine is distinct from the "globule concepts" John Pickstone has described from the early nineteenth century in "Globules and Coagula: Concepts of Tissue Formation in the Early Nineteenth Century," *Journal of the History of Medicine and Allied Sciences* 28: 336–356 (1973). The "globules" described up until the late 1820s were not real but artifacts produced by chromatic and spherical aberration before the development of achromatic lenses. The "molecules" of Schwann, Bennett, and others were much smaller but could be clearly seen with the new achromatic lenses, as well as by Needham, Buffon, Brown, and others with excellent single-lens microscopes.
6. One historian who does make note of the doctrine of histological molecules is Leland Rather, in *Addison and the White Corpuscles*, pp. 25–30, 34.
7. Richard Owen, *On the Anatomy of Vertebrates*, v. 3 (London: Longmans, Green, 1868), pp. 816–817.
8. Philip Sloan, "Darwin's Invertebrate Program, 1826–1836: Preconditions for Transformism," pp. 71–120 in David Kohn, ed., *The Darwinian Heritage* (Princeton: Princeton University Press, 1985); Sloan, "Darwin, Vital Matter and the Transformism of Species," *Journal of the History of Biology* 19: 369–445 (1986); Rather, *Addison and the White Corpuscles*, see esp.

p. 80. Addison had accepted that most of Brown's "active molecules" were moved by physical-chemical forces. He distinguished his own histological molecules, however, saying that their movement was entirely vital. He felt it was of interest "to determine whether they originate from the active molecules off the mucous cells [Schwann-like, tantamount to heterogenesis for many observers] . . . or from invisible ova in the atmosphere; or from both sources." Thus the role of "molecules" in generation, spontaneous or otherwise, was still very much up for grabs in 1844. Taken overall, however, Addison's view seems to be somewhat vitalistic, so that his molecules are very much like Beale's particles of "bioplasm" (1870). The irony, of course, is that just as Schwann opposed spontaneous generation and did not see how similar his ideas were to heterogenesis, so Beale's particles are sufficiently similar to Pouchet's vitalistic heterogenesis that many thought Beale was confirming Pouchet's views. See Crellin, "Dawn of the Germ Theory," p. 70. Crellin notes that the same can be said of Huxley's views on "Xenogenesis."

9. William Coleman, *Biology in the Nineteenth Century* (Cambridge: Cambridge University Press, 1971); L. S. Jacyna, "John Goodsir and the Making of Cellular Reality," *Journal of the History of Biology* 16: 75–99 (1983); Jacyna, "The Romantic Programme and the Reception of Cell Theory in Britain," *Journal of the History of Biology* 17: 13–48 (1984). For Endlicher and Unger, see Sander Gliboff, "Evolution, Revolution, and Reform in Vienna: Franz Unger's Ideas on Descent and their Post-1848 Reception," *Journal of the History of Biology* 31: 179–209 (1998), pp. 186–189.
10. John Farley (*The Spontaneous Generation Controversy from Descartes to Oparin* (Baltimore: Johns Hopkins University Press, 1977), p. 52), for example, seems to completely miss the fact that Schwann's "molecules" were seen as histological molecules by most readers (and probably intended as such by Schwann), *not* chemical molecules. Similarly, Henry Harris overlooks the connection of the "molecules" to spontaneous generation ideas in his *The Birth of the Cell* (New Haven: Yale University Press, 1999), pp. 52, 100–101.
11. Sloan, "Darwin's Invertebrate Program"; Sloan, "Darwin, Vital Matter."
12. Charles Darwin, *The Variation of Animals and Plants Under Domestication*, 2 vols. (London: John Murray, 1868), v. 2, pp. 357–404; Owen, *Anatomy of the Vertebrates*, v. 3, pp. 810–824.
13. Darwin, *Variation of Animals and Plants*, pp. 370–371.
14. Owen, *Anatomy of the Vertebrates*, p. 812. Owen's likening of the Pasteur-Pouchet debate to the Cuvier-Geoffroy debate of 1830 has been noted by John Farley and Gerald Geison in "Science, Politics and Spontaneous Generation in Nineteenth Century France: The Pasteur-Pouchet Debate," *Bul-*

letin of the History of Medicine 48: 161–198 (1974), see p. 167; and more recently by Evelleen Richards ("A Political Anatomy of Monsters, Hopeful and Otherwise," *Isis* 85: 377–411 (1994), p. 392). The irony has not been noted, however, that Owen thus saw Darwin and Pasteur as bedfellows, in a way that Pasteur and most orthodox French science under the Second Empire would (in Farley and Geison's account) have found *very* objectionable. Owen noted that despite oratorical skills inferior to those of his opponent, "Pouchet is rapidly acquiring, in reference to the origin of monads, that position which Geoffroy Saint-Hilaire has taken in regard to the origin of species" (Owen, *Anatomy of the Vertebrates*, v. 3, p. 814).

15. Evelleen Richards, "A Question of Property Rights: Richard Owen's Evolutionism Reassessed," *British Journal for the History of Science* 20: 129–171 (1987), esp. pp. 155–157 and Richards, "A Political Anatomy of Monsters," esp. pp. 392–400.
16. Nicolaas Rupke, *Richard Owen: Victorian Naturalist* (New Haven: Yale University Press, 1994), pp. 240–245, 250–252.
17. See Robert Grant, *Recent Zoology: A Tabular View of the Primary Divisions of the Animal Kingdom* (London: Walton and Maberly, 1861), pp. 5–6, 9.
18. Sloan, "Darwin, Vital Matter," p. 429.
19. Sloan, "Darwin's Invertebrate Program."
20. Sloan, "Darwin, Vital Matter," p. 434.
21. Sloan, "Organic Molecules Revisited," in *Buffon '88* (Paris: Vrin, 1992), pp. 415–438.
22. Adrian Desmond and James R. Moore, *Darwin* (London: Michael Joseph, 1991) pp. 82, 223.
23. Anon., "Obituary of Robert Brown," *Athenaeum* 1599: 786 (19 June 1858). On the Continent the term used was "Brunonian movement."
24. Sloan, "Darwin, Vital Matter," p. 435.
25. Christian Ehrenberg, "Über das entstehen des organischen aus einfacher sichtbarer materie," *Annalen der Physik und Chemie* 24(n.s.): 1–48 (1832). English trans. appeared in *Scientific Memoirs*, v. 1, ed. Richard Taylor, pp. 555–583 (London: R. and J. Taylor, 1837).
26. Sloan, "Darwin, Vital Matter," p. 436, emphasis in original. Ehrenberg's paper addressed the larger question of the existence of "atoms," or simplest living organic units, and he directly addressed Brown's observations in this light. Ehrenberg, like Brown, saw the granular particles as possessing "lebendigen Bewegungen," though he was anxious to show in his discussion that this did not imply spontaneous generation, as Buffon inferred from his "organic molecules."
27. On the continuity of Darwin's thought over decades on pangenesis, see Francis Darwin, ed., *The Autobiography of Charles Darwin and Selected Let-*

ters (New York: Appleton, 1892), p. 281; also Robert Olby, "Charles Darwin's Manuscript of Pangenesis," *British Journal for the History of Science* 1: 251–263 (1963).

28. See L. Oken, *Lehrbuch der Naturphilosophie,* 1809, pars. 901–956 (pp. 185–192 in the first English trans., *Elements of Physio-philosophy* [London: Ray Society, 1847]) for Oken's ideas on spontaneous generation.
29. Michael Ruse suggests that Owen's *public* opposition to philosophical anatomy and *Naturphilosophie* during this period should not be taken entirely at face value, since Owen even more than Darwin was dependent on his conservative patrons for support and knew what they wanted to hear. See "Were Owen and Darwin *Naturphilosophen?*" *Annals of Science* 50: 383–388 (1993), p. 386.
30. See Toby Appel, *The Cuvier-Geoffroy Debate* (Oxford: Oxford University Press, 1987) on the reevaluation and revival of Geoffroy's reputation after 1830. On the resemblance of Owen's work to Oken's by 1852 or so, see *LLTHH,* v. 1, p. 112.
31. Adrian Desmond has shown that crucial to demonstrating his loyalty to his conservative patrons, Owen felt the need to respond to agitation by the radical medical community, especially Robert Grant. Desmond has also shown that this was crucial to Owen's fortification of an institutional power base, largely at the direct expense of Grant, his immediate rival at the Zoological Society of London. See "Robert E. Grant: The Predicament of a Pre-Darwinian Transmutationist," *Journal of the History of Biology* 17: 189–223 (1984), esp. pp. 219–221.
32. James Secord, "Extraordinary Experiment: Electricity and the Creation of Life in Victorian England," pp. 337–383 in David Gooding, Trevor Pinch, and Simon Schaffer, eds., *The Uses of Experiment* (Cambridge: Cambridge University Press, 1989), quotation on p. 338.
33. Crosse to Owen, Owen papers BMNH 9: 108–113. See also Jacob Gruber, "The Richard Owen Correspondence: An Annotated Calendar," unpublished ms., APS Library, Philadelphia.
34. See Oliver Stallybrass, "How Faraday 'Produced Living Animalculae': Andrew Crosse and the Story of a Myth," *Proceedings of the Royal Institution* 41: 597–619 (1967).
35. Richard Owen, *The Hunterian Lectures in Comparative Anatomy, May-June 1837,* ed. P. R. Sloan (Chicago: University of Chicago Press, 1992), p. 107.
36. See the ms. notes for Owen's 1840 Hunterian Lectures, Owen papers, BMNH v. 38(1).
37. Ibid., ff 9–11 of final copy of lecture #2, 23 Apr. 1840 (though apparently not actually delivered until 4/30). Thomson's successor in the Edinburgh Chair of the Institutes of Medicine, John Hughes Bennett, seems also to

have taken Thomson's article seriously, as I will discuss later. Burdach and Oken are discussed on ff. 3, 7–8, 13–15. Owen takes up the question of spontaneous generation of intestinal worms on f. 36.

38. See Oken, *Lehrbuch,* pars. 901–956; pp. 185–192 of first English trans., *Elements of Physio-philosophy.*
39. Richards, "A Question of Property Rights," pp. 150–167.
40. Theodor Schwann, *Microscopische Untersuchungen* (1839), English trans. by H. Smith as *Microscopical Researches into the Accordance in the Structure and Growth of Animals and Plants* (London: Sydenham Society, 1847), pp. 191–197.
41. L. S. Jacyna, "John Goodsir and the Making of Cellular Reality," *Journal of the History of Biology* 16: 75–99 (1983), pp. 75–76.
42. Rather, *Addison and the White Corpuscles,* pp. 25–26, quotation from p. 43.
43. Ibid., pp. 218–219.
44. Ibid.
45. Bennett ms. lecture notes, "On the Doctrine of Life and the Tissues which Manifest It," Box Gen. 2007/1, Bennett papers, Edinburgh University Library. For more on Bennett and his background, see L. S. Jacyna, "John Hughes Bennett and the Origins of Medical Microscopy in Edinburgh: Lilliputian Wonders," *Proceedings of the Royal College of Physicians of Edinburgh* 27 (Suppl. 3): 12–21 (1997).
46. Ibid. Bennett was critiquing Schwann, *Microscopical Researches,* pp. 191–195, 201–210. Bennett refers to John Snow's *On Continuous Molecular Changes* in this ms. (p. 34) as well. See also Bennett's first published version of his "molecular theory" in his article, "Physiology," v. 17 (1859): 648–703, 8th ed., *Encyclopedia Britannica,* pp. 648–651. For other British critics on Huxley's 1853 article, see Marsha Richmond, "T. H. Huxley's Criticism of German Cell Theory: An Epigenetic and Physiological Interpretation of Cell Structure," *Journal of the History of Biology* 33 (forthcoming).
47. John Hughes Bennett, "Lectures on Molecular Physiology, Pathology and Therapeutics," *Lancet* 1: 1–4 (3 Jan. 1863), quotation from p. 1.
48. Ibid.
49. Rather, *The Genesis of Cancer,* pp. 67–68. These particles also seem to be what Wilhelm Reich called "bions" a century later. See Reich, *Die Bione,* English trans. by Derek and Inge Jordan as *The Bion Experiments* (New York: Farrar, Straus, and Giroux, 1979). Even T. H. Huxley in his 1870 BAAS presidential address seems to still believe this type of heterogenesis is a reality (he suggests renaming it "Xenogenesis"), despite his insistence otherwise that any exceptions to "homogenetic biogenesis" (life only from like parents) are probably illusory. See *Nature* 2: 404 (15 Sept. 1870) or *Addresses Biological and Geological,* pp. 261–262.

50. Addison, "On Moving Molecules in the Interior of Cells," *Provincial Medical and Surgical Journal* 7: 137–138 (1844); see also anon. review of Addison in *British and Foreign Medical Review* 17: 92–96 (1844). An extended discussion of Addison's ideas and his interaction with Bennett can be found in Rather, *Addison and the White Corpuscles,* esp. pp. 25–31, 34, 39–48, 70–91, 217–221.
51. Bennett, "On the Molecular Theory of Organization, read 1st April 1861" *Proceedings of the Royal Society of Edinburgh* 4: 436–446 (1857–1862), quotation on p. 446. Note that in the German context not only Oken, but also former Schleiden student Karl von Nägeli had made the mental connection between spontaneous generation and "molecules" earlier than most in Britain (with the exception of Robert Grant). Pauline Mazumdar has discussed Nägeli's conversion and suggested that it represents the logical extension of Schleiden's "Unitarian" view of life and nature. See her *Species and Specificity* (Cambridge: Cambridge University Press, 1995), esp. pp. 26–34, 37–40. Nägeli's work in 1852–1855 on the structure of starch grains from simple, microscopically visible organic units was important in his conversion to belief in spontaneous generation. Mazumdar says "it seems probable that he did so as a contemporary and critic of Schwann's and particularly of Schleiden's cell theories, for which the growth of the starch grains modeled that of the growth of cells themselves, at the transition between the organic and the inorganic worlds" (p. 32). Interestingly, Nägeli made this transition at the same time that his friend Albert Kölliker moved, with Virchow, in the opposite direction, away from the cytoblastema hypothesis. See also J. S. Wilkie, "Nägeli's Work on the Fine Structure of Living Matter," *Annals of Science* 16: 11–41, 171–207, 209–239 (1960) and 17: 27–62 (1961).
52. Bennett, "Lectures on Molecular Physiology," p. 1.
53. Ibid., p. 57.
54. In addition to his scrap with Virchow over priority in describing leukemia, Bennett was at the time immersed in a heated controversy on the uselessness (in his view) of bloodletting. On this see John Harley Warner, *The Therapeutic Perspective* (Cambridge, Mass.: Harvard University Press, 1986), pp. 215–224.
55. Bennett, "On the Molecular Theory of Organization," quotation on p. 436.
56. Anon., "Review of Virchow's *Cellular Pathology,*" *BMJ* 1: 44–46, 94–96 (12 Jan. 1861), 94–96; also v. 2: 378 (7 Oct. 1865). The reviewer also accused Virchow of plagiarizing Goodsir's idea of "cell-territories" and "cells as centers of reproduction of other cells," saying he "derived his leading views as to cells from one Edinburgh professor, so he has borrowed, and equally without acknowledgment, his notions concerning leukaemia from another, viz., Professor Bennett" (p. 44).

57. Bennett, "Lectures on Molecular Physiology," pp. 1–4, 55–57, 139–141, 199–202, 259–262, 378–381, 459–461, 597–600; v. 2: 3–6, 239–242, 643–646, and 671–675 (1863).
58. Lionel Beale, "Facts and Arguments Opposed to Dr. Bennett's Theory of Organization," *BMJ* 1: 109–110, 157–158, 235–236, 365–366 (1863).
59. John Hughes Bennett, "Remarks on the Molecular Theory of Organization: in Reply to Dr. Beale," *BMJ* 1: 135–136, 209–210, 313–314 (1863). Note that after Bennett declared he had discredited Beale's claims and withdrew (28 March 1863), Beale got in one more response and, thus, the last word.
60. Beale, "Facts and Arguments," p. 109.
61. See the version of the molecular theory in Bennett's *Clinical Lectures on the Principles and Practice of Medicine,* 4th English (Edinburgh, 1865) & 5th American ed. (New York: Wood & Co., 1867), pp. 118–124.
62. See John Hughes Bennett, "Atmospheric Germ Theory: A Lecture Delivered to the Royal College of Surgeons in Edinburgh, January 17, 1868," *Edinburgh Medical Journal* 13: 810–834 (1868), pp. 830–834; and see the more pointed singling out of Beale's vitalism in Bennett's *Textbook of Physiology* (Philadelphia: Lippincott, 1872), pp. 55–56, 176–186.
63. On Beale's exchange with Owen, see *MMJ* 1: 178–180, 294–295 (1869); on Beale's more protracted battle with Huxley, see Gerald Geison, "The Protoplasmic Theory of Life and The Vitalist/Mechanist Debate," *Isis* 60: 273–292 (1969).
64. Bennett, "Atmospheric Germ Theory," pp. 812–813.
65. Ibid., pp. 813n, 818. This charge was quickly echoed by other British heterogenists such as Gilbert Child [*PRSL* 1864: 184–185; *Essays on Physiological Subjects* (London: Longmans, 1869), pp. 129–130] and Owen (*Anatomy of the Vertebrates* 3: 815).
66. Beale, "Facts and Arguments," p. 109.
67. Bennett, "Remarks . . . in Reply to Beale," p. 135.
68. See John Snow, *On Continuous Molecular Changes* (London, 1853), and numerous others described in M. Pelling, *Cholera, Fever, and English Medicine, 1825–1865* (Oxford: Oxford University Press, 1979).
69. Richard Owen, *Palaeontology* (Edinburgh: Black, 1860), p. 441; Owen, "On the Aye Aye," *Transactions of the Zoological Society of London* 5: 33–101 (1862–1865), p. 92. This paper had been read to the Zoological Society on 14 and 28 Jan. 1862. Owen's language on heterogenesis had already begun to hedge on support for the doctrine as early as his Aug. 1858 BAAS presidential address.
70. Indeed, Charles Lyell also guessed as much in a letter to Darwin on 15 Mar. 1863: "As to Owen in his Aye Aye paper, he seems to me a disciple of Pouchet, who converted him at Rouen to 'spontaneous generation'" *LLCL,*

p. 366; published in *Correspondence of Charles Darwin,* v. 11 (Cambridge: Cambridge University Press, 2000), pp. 230–231 (*DCC* #4041).

71. Pouchet to Owen, BMNH 21: 417–419.
72. Owen, "On the Aye Aye," p. 92.
73. Wyman to Owen, Owen papers, BMNH 27: 254.
74. [Owen], "Review of Carpenter's *Introduction to the Study of the Foraminifera,*" *Athenaeum* 28 Mar. 1863: 417–419, quotation from pp. 418–419. Owen's review reprinted in *Correspondence of Darwin,* v. 11, pp. 754–765.
75. The analogy used in this review between an amoeba and a magnet reappears almost unchanged, for example, in Owen's fully developed defense of spontaneous generation in 1868.
76. W. B. Carpenter, "Dr. Carpenter and His Reviewer," *Athenaeum* 4 Apr. 1863: 461; *Correspondence of Darwin,* v. 11, pp. 765–767.
77. Charles Darwin, "The Doctrine of Heterogeny and Modification of Species," *Athenaeum* 25 Apr. 1863: 554–555, quotation on p. 554. Darwin had been chafing at Owen's attacks and just waiting for a chance to get a shot off. See his letter to Lyell of 17 Mar. 1863 (*Correspondence of Darwin,* v. 11, pp. 243–244; *DCC* #4047): "I have read Owen's 'Aye Aye'; I could find nothing to lay hold of, which in one sense . . . I am very sorry for, as I long to be in the same boat with all my friends and I had written so good a letter, all ready, with a blank for his sentence claiming more than he had any right to; but I could pick out no such sentence."
78. Darwin to Hooker 29 Mar. 1863, Darwin to Hooker correspondence vol. p. 732, Hooker papers, Royal Botanical Gardens, Kew; published in *Correspondence of Darwin,* v. 11, pp. 277–278 (*DCC* #4065); expurgated version in Darwin's *Autobiography,* p. 272 and in *LLCD,* v. 3: 17–18. It is interesting to note the 1847 translator's choice of "mucus" as a rendering of Oken's original "schleim" for the primordial substance from which all living matter derives. By the 1860s this would have been rendered as "sarcode" or "protoplasm," since at that time Haeckel reconstructed Huxley's *Bathybius* protoplasm as the fulfillment of Oken's predicted "Urschleim" from the sea floor. Indeed in his 1863 review of Carpenter, Owen wrote "the difference between snot and sarcode, between passive and active states of mucus, is due to a conversion of attractive or other mode of polar force into the mode called 'assimilative,' 'organic,' or 'vital'" (pp. 418–419). This is the passage Darwin refers to in his letter to Hooker.
79. Darwin to Bentham 22 May 1863, Bentham correspondence v. 3: 711, Hooker papers, RBG Kew; published in *Correspondence of Darwin,* v. 11, pp. 432–433 (*DCC* #4176).
80. Editorial comment by Francis Darwin, *LLCD* 3: 24. My thanks to Marsha

Richmond and the staff of the Darwin Letters Project for clarifying that was comment was actually by Francis Darwin rather than his father.

81. Hooker to Darwin, 7 May 1863, *LLJDH* 2: 51; published in *Correspondence of Darwin,* v. 11, pp. 387–388 (*DCC* #4144). Another public disagreement that had gained much notoriety just prior to this resulted from the use of public forums by Owen's client Paul DuChaillu.
82. Darwin to Hooker, [9 May 1863], *Correspondence of Darwin,* v. 11, pp. 393–395 (*DCC* #4148); excerpt in Darwin's *Autobiography,* p. 273.
83. See Christopher Hamlin, "Scientific Method and Expert Witnessing: Victorian Perspectives on a Modern Problem," *Social Studies of Science* 16: 485–513 (1986).
84. Huxley to Kingsley 22 May 1863, *LLTHH,* 1: 263.
85. [Child], "Recent Researches on the Production of Infusoria," *BFMCR* 34: 102–123 (1864). Child, *Essays on Physiological Subjects* (London: Longmans, 1868), 2nd enl. ed., 1869.
86. Anon., "Report on the 39th Annual Meeting of the Association of German Naturalists and Physicians," *BMJ* 2: 495 (1864). Owen had clipped a copy of Schaaffhausen's published paper and kept it in his papers on generation [BMNH, Owen papers, 90 (3)]. Attached was a ms. summary of Owen's ideas on spontaneous generation at around this time. For more on Schaaffhausen and his support of Darwinism, see E. Roth, "Obituary of Hermann Schaaffhausen," *Leopoldina* 29: 168–173, 185–189 (1893).
87. Jeffries Wyman was a long-standing friendly correspondent of Owen. But his spontaneous generation experiments were also respected and taken seriously by Darwin. He makes an interesting case study for the disputed areas between Darwin and Owen and how one might bridge those disputes if not involved in the past personal bitterness between the two men. See Toby Appel, "Jeffries Wyman, Philosophical Anatomy, and the Scientific Reception of Darwin in America," *Journal of the History of Biology* 21: 69–94 (1988).
88. Anon., "The Spontaneous Generation Question," *Lancet* 1: 234 (13 Feb. 1869).
89. On both the politically radical medical movement and Owen's staunch, even armed, opposition to it in the 1830s and 1840s, see A. Desmond, *The Politics of Evolution* (Chicago: University of Chicago Press, 1989). The *Lancet* had already reaffirmed its support for spontaneous generation in 1864 by taking a position in defense of Pouchet: "Heterogenesis," *Lancet* 2: 666 (10 Dec. 1864).
90. Anon., "The Spontaneous Generation Question," p. 234.
91. [H. C. Bastian], "The Origin of Life," part vii, *BMJ* 2: 665–666 (18 Dec. 1869), p. 665.
92. Owen, *Anatomy of Vertebrates,* pp. 815–817.

93. Bennett, "Atmospheric Germ Theory." Reprinted in shortened form as "On the Molecular Origin of Infusoria," *Popular Science Review* 8: 51–66 (1869). This was cited by Gilbert Child as well, in both the 1868 and 1869 editions of his *Essays on Physiological Subjects,* where he also argued forcefully for spontaneous generation as a necessary corollary of Darwinism.
94. Wyman to Burt Wilder, Mar. 1865, Wilder papers, Box 1, Cornell University Library.
95. M. Faraday, "On Lines of Magnetic Force; Their Definite Character, and Their Distribution Within a Magnet and Through Space," *PTRSL* 142: 25–56 (1852). Faraday's ideas on matter as related to fields or lines of force have also been previously attributed to the influence of *Naturphilosophie,* through the work of Oersted. See L. Pearce Williams, *The Origins of Field Theory* (New York: University Press of America, 1980), pp. 51–63.
96. Owen, *Anatomy of Vertebrates,* p. 819.
97. Ibid., p. 822.
98. Ibid.
99. On the role of Mesmerists and the opposition of Faraday and Carpenter, see Alison Winter, "The Construction of Orthodoxies and Heterodoxies in the Early Victorian Life Sciences," in Bernard Lightman, ed., *Victorian Science in Context* (Chicago: University of Chicago Press, 1997); also Winter, *Mesmerized: Powers of Mind in Victorian Britain* (Chicago: University of Chicago Press, 1998). On the "electricians," including Crosse, see Iwan Rhys Morus, "Currents from the Underworld: Electricity and the Technology of Display in Early Victorian England," *Isis* 84: 50–69 (1993). Also Morus, *The Children of Frankenstein* (Princeton: Princeton University Press, 1998). Interestingly, one of the few attempts to revive Crosse's reputation in the early twentieth century was by Jerome Alexander, the colloid chemist interested in the origin of life. See "Andrew Crosse: Electrical Pioneer," *Nature* 134: 105 (1934).
100. Child, *Essays on Physiological Subjects,* 2nd ed., pp. 141–143.
101. Owen to Bastian, 6 June 1880, Owen papers (B Ow 2.10), APS Library, Philadelphia.
102. Bastian to Owen, 28 Oct. 1876, Owen papers BMNH 2: 324–325.
103. Charles Letourneau, *Biology,* English trans. by W. Maccall (London: Chapman and Hall, 1878), pp. 341–342.
104. Brown papers, BMNH (Botany lib.), Solander box 24, file 70 (Onagrariæ), ff. 224v, 13 June 1827. On 4 July 1827, George Bentham recorded that "Brown, under a power of 600 diameters, showed him the 'animalculae of pollen' from *Lolium perenne.*" See H. D. Jackson, *George Bentham* (London: J. M. Dent, 1906), p. 62.
105. Darwin, *Variation of Animals and Plants,* v. 2, pp. 370–371. Darwin cites Edmund Montgomery as an example of the blastema approach rather than

Bennett, and his lack of familiarity with the histological literature makes it unclear whether he had read Bennett's work up to this point, but his remark would clearly apply perfectly well to Bennett's molecular theory. Darwin had consulted Edinburgh physiologist and Bennett supporter William Turner on the current state of cytoblastema views. See Darwin to Turner, 5 June 1866, Darwin papers, APS Library (*DCC* #5113).

106. Bennett, "Atmospheric Germ Theory." Note that Bennett's paper was read publicly just before the publication of Darwin's book, but did not appear in print until two months later, in the Mar. 1868 *Edinburgh Medical Journal.*
107. Also included among these writers were Gilbert Child of Oxford University; Jeffries Wyman of Harvard University; Henry Lawson of St. Mary's Hospital Medical School; Richard Owen; Arthur Durham, president of the Quekett Microscopical Club; and Francis Abbott, a reviewer of Spencer's *Principles of Biology* for the *North American Review.*
108. Through the early 1860s even a figure of the stature of Richard Owen had trouble getting heterogenesis a fair hearing (1858 Leeds address, 1860 *Palaeontology,* "Aye Aye" paper), and thus began to fly his speculations along these lines as trial balloons only in anonymous reviews (e.g., *Athenæum* 28 Mar. and 2 May 1863). Here I differ with Rupke (*Richard Owen: Victorian Naturalist*) in believing Owen's commitment to these theories was more than merely a strategy for differentiating his evolutionary ideas from those of Darwin.
109. Bennett, "Atmospheric Germ Theory," p. 829n.
110. James Samuelson, "On the Source of Living Organisms," *Quarterly Journal of Science* 1: 598–614 (1864), p. 606n. In this paper Samuelson also expresses admiration for Child's arguments of 1864 in the *BFMCR.*
111. A. Thomson, "Generation," in *Todd's Cyclopedia of Anatomy and Physiology* (London: Longman, 1839), v. 2: 431. On the wide reputation of this book, see W. J. O'Connor, *Founders of British Physiology* (Manchester: Manchester University Press, 1988), p. 36–39. Richard Owen discussed the article at length in his Hunterian Lectures of 1840, Owen papers, BMNH 38 (1), lecture #2, final copy, pp. 10–12.
112. L. Stephen Jacyna, *Philosophic Whigs* (London: Routledge, 1994).
113. Owen, Hunterian Lectures, ms. lecture #2, p. 10.
114. See Darwin, *Variation of Animals and Plants,* v. 2, pp. 378–382; Child, *Essays on Physiological Subjects,* 2nd ed., pp. 148–155; on Farr, see J. M. Eyler, *Victorian Social Medicine* (Baltimore: Johns Hopkins, 1979), esp. pp. 105–108; on the profusion of particulate theories at this time, see also Crellin, "Dawn of the Germ Theory," pp. 57–76. Crellin fails, however, to include Graham's colloids in this category, where I will show it was an important member.

115. See Pelling, *Cholera, Fever, and English Medicine,* esp. pp. 301–304 on how the British medical community did *not* see Liebig's and Pasteur's views as irreconcilable. Pelling also shows that Bennett was among a great many British medical men who looked askance at Pasteurian-style germ theories because of their memory of the failed cholera-fungus theory of 1849. Bennett discusses this in "Atmospheric Germ Theory," p. 829. For another good example at just this time, see Henry Bence Jones, *Lectures on Applications of Chemistry and Mechanics to Pathology and Therapeutics* (London: John Churchill, 1867), pp. 221–222.
116. Child, *Essays on Physiological Subjects,* p. 139.
117. Philip Rehbock, "Huxley, Haeckel and the Oceanographers: The Case of *Bathybius haeckelii,*" *Isis* 66: 504–533 (1975), see esp. pp. 516–524.
118. George Wallich, "On the Vital Functions of the Deep-Sea Protozoa," *MMJ* 1: 32–41 (1 Jan. 1869); also Wallich, "On the Rhizopoda as Embodying the Primordial Type of Animal Life," *MMJ* 1: 228–235 (1 Apr. 1869).
119. Ernst Haeckel, "Beiträge zur Plastidientheorie," *Jenaische Zeitschrift* 5: 499–519 (1870), p. 500, quoted in F. Lange, *History of Materialism,* v. 2: 20n. Indeed, E. Ray Lankester, Huxley's protégé, was so taken with Haeckel's ideas that he supervised the first English translation of Haeckel's book *Natürliche Schöpfungsgeschichte,* enthusing over *Bathybius,* in 1874. Unfortunately for him, the book did not appear until 1876, *after* Huxley's embarrassed public retraction in 1875 when *Bathybius* was discovered to be an artifactual chemical precipitate.
120. Friedrich Lange, *The History of Materialism,* English trans. by E. C. Thomas (New York: Harcourt Brace and Co., 1877), 2: 19.
121. Ibid., p. 14.
122. Ibid., pp. 24–25.
123. Darwin, by 1870, believed that gemmules from the cells of infusoria might survive boiling and other hostile conditions in spontaneous generation experiments, and thus serve as the source of new microorganisms that afterwards appeared in the infusions. See his letter to Jos. Hooker of 12 July 1870, *MLCD,* 1: 321–322 (*DCC* #7273).
124. Henry Lawson, "Prof. Owen's Anatomy," *Popular Science Review* 8: 67–71 (1869). Ironically, the Darwinians closest to Huxley could see no two texts as more different than Huxley's *Lectures on the Elements of Comparative Anatomy* and Owen's *Anatomy of the Vertebrates.* Michael Foster, in an Oct. 1869 review of both books, joked about how extreme were the personal and theoretical differences between the two authors:

    > When in reading [Huxley] we are startled in the midst of a discussion from which technical language has filtered off any excess of excitement, by the sudden dash of a piercing sarcasm which makes a man wince even

> as it passes by him, we know at once whither the winged words were aimed. And when in [Owen] at unexpected corners and in culs-de-sac of notes we come upon heavy reproaches hurled against certain Roman numerals, we have no need to turn to the end of the volume to learn the name to which those numerals refer . . . It is hardly too much to say that were the authors under an agreement to write in common a work on The Principles of Biology, in the form of a series of definitions or abstract propositions, they would not be able to put down on paper more than the bare title.

Foster, "Higher and Lower Animals," *Quarterly Review* 127: 381–400 (1869), p. 382.

125. Lawson, "Professor Owen's Anatomy," quotations on pp. 70–71. Lawson also began as editor of a new technical periodical, *The Monthly Microscopical Journal,* in that year, in whose pages much of the debate took place over the next few years.
126. T. R. R. Stebbing, "On Spontaneous Generation," in his *Essays on Darwinism* (London: 1871).
127. H. Holland, "Life on the Earth," in his *Fragmentary Papers* (London: 1875), esp. pp. 129–134. Holland had apparently reconsidered his remark of April 1862 when Darwin first referred him to Pasteur's 1861 memoir in response to an earlier draft of this essay. At that time Holland had said: "Pasteur's memoir, to which you allude, is a very able and convincing one. He completely pushes Pouchet from the field" (*DCC* #3490; published in *Correspondence of Darwin,* v. 10, pp. 141–142).
128. See G. L. Geison, "The Protoplasmic Theory of Life and the Vitalist-Mechanist Debate," *Isis* 60: 273–292 (1969).
129. Ibid. Beale's initial salvo of 14 Apr. 1869 was published as "Protoplasm and Living Matter," *MMJ* 1: 277–288 (1869).
130. L. S. Beale, "Professor Owen on Magnetic and Amoebal Phenomena," *MMJ* 1: 178–180 (1 Mar. 1869).
131. Ibid., p. 178.
132. Ibid., pp. 178–179, emphasis in original.
133. R. Owen, "Professor Owen on Article VI, No. III, of the '*Monthly Microscopical Journal,*'" *MMJ* 1: 294–295 (1 May 1869), p. 294.
134. Ibid., p. 295.
135. J. H. Stirling, *As Regards Protoplasm* (Edinburgh: Blackwood and Sons, 1869), 2nd ed. 1872, p. 67, italics in original. First presented as a lecture to the Royal College of Physicians of Edinburgh on 28 Apr. 1869. As late as 1880 John Ruskin again referred mockingly to the Crosse experiments in a widely read article, showing that Crosse's acari were still a bugbear even then, especially in the hands of critics of Huxley, Tyndall, and their ilk. See

Ruskin, "Fiction, Fair and Foul," *Nineteenth Century* 7: 941–965 (June 1880), pp. 943–944. The use of Crosse's acari as a foil for any legitimate science on origin of life remained very much alive in Britain in 1952; see N. W. Pirie, "Vital Blarney," *New Biology* 12: 106–112, p. 110.

136. Stirling, *As Regards Protoplasm*, p. 69.
137. Ibid., pp. 50–51.

3 Bastian as Rising Star

1. Edward L. Youmans, "Sketch of Dr. H. C. Bastian," *Popular Science Monthly* 8: 108–110 (1875–76), p. 109. More on Bastian can be found in his *DNB* entries: John Walton of Detchant, in *DNB* ("Missing Persons" vol.), p. 47; and Mick Worboys entry in the new version of the *DNB*, currently in preparation.
2. Mercer Rang, "Henry Charlton Bastian, 1837–1915," *University College Hospital Magazine* 39: 68–73 (1954), p. 68.
3. Adrian Desmond and James R. Moore, *Darwin* (London: Michael Joseph, 1991) pp. 594–595.
4. Since the *Darwin* biography, Desmond has moved a considerable way in the direction of my interpretation, first argued in my Jan. 1997 Ph.D. dissertation. See Adrian Desmond, *Huxley: From Devil's Disciple to Evolution's High Priest* (Reading, Mass.: Addison-Wesley, 1997), pp. 392–393.
5. Unlike Darwin, Bastian did not break off all contact with Grant as he rose in scientific stature. Extant letters between the two are cordial in tone, and Grant was clearly pleased with Bastian's success in winning respectability for spontaneous generation. See Grant to Bastian, 26 June 1872 and 10 Feb. 1873, Wellcome Institute London, autograph letters collection. However, Bastian seemed quite as aware as Darwin that too close an association with Grant would not help his cause. Even in his huge *Beginnings of Life*, practically an exhaustive catalog of spontaneous generation support up to 1872, Bastian mentions Grant only briefly and in no way suggests he is Grant's disciple.
6. Mercer Rang, *The Life and Work of Henry Charlton Bastian*, unpublished ms., 1954, in University College Medical Library, p. 4; Bastian to (Sharpey?), 6 Aug. 1863, University College Library, College Correspondence file, unsorted; Youmans, "Sketch," p. 109.
7. Sir John Meyer to Bastian, 5 Jan. 1866, Wellcome Institute London, autograph letters collection.
8. Youmans, "Sketch," p. 109.
9. Zachary Cope, *The History of St. Mary's Hospital Medical School* (London: Wm. Heinemann, 1954), pp. 85, 124, 139.

10. Youmans, "Sketch," p. 110.
11. Cope, *History of St. Mary's*, pp. 168–169, 178. Hart became editor of the *BMJ* in 1866 and continued until his death in 1898. "He did much to increase the influence of the B. M. A. and also helped the cause of sanitary reform." From 1868–1870, Lawson wrote strongly favorable reviews of evolution and spontaneous generation (which he saw as linked) as editor of *MMJ, Popular Science Review*, and *Scientific Opinion*. See W. H. Brock, "Patronage and Publishing: Journals of Microscopy, 1839–1989," *Journal of Microscopy* 155: 249–266 (1989), p. 253 on Lawson's editorial activities and on the *MMJ* as attempting to fulfill Huxley and Flower's vision of a journal. He was also co-editor, from 1 July 1868 to 30 June 1869, of the *Practitioner.* However, sometime in 1871–72 Lawson became converted to the Huxley/Tyndall opposition and afterwards, until his untimely death in Oct. 1877 at age thirty-seven, lent his authority to opposing Bastian and promoting Bastian's opponents, especially William Dallinger.
12. Francis Sibson to Grayson, 28 Apr. 1866, Wellcome Institute London, autograph letters collection. Sibson had been elected FRS in 1849, "chiefly for his accurate and painstaking work on the anatomy of the viscera in health and disease." See Cope, *History of St. Mary's*, pp. 209–210.
13. Huxley Reviewers Report, RR.6.18, RS Archives.
14. Busk Rev. Report, RR.6.37, RS Archives.
15. Bastian, "On the Anatomy and Physiology of the Nematoids, Parasitic and Free," *PTRSL* 156: 545–624 (1866).
16. Glaishier was president of the Royal Microscopical Society from 1866–1870. See G. L'E. Turner, "The Origins of the Royal Microscopical Society," *Journal of Microscopy* 155: 235–248 (1989), pp. 246–247.
17. "Certificate of a Candidate for Election for H. C. Bastian," RS Archives.
18. Roy MacLeod, "Seeds of Competition," *Nature* 224: 431–434 (1969), p. 431.
19. Gordon Holmes, *The National Hospital Queen Square, 1860–1948* (London: E. & S. Livingstone, 1954), p. 39.
20. Bastian to UCL, 15 Nov. 1867, UCL Library, College Correspondence Collection, unsorted.
21. Holmes, *National Hospital*, pp. 38–39. On Russell Reynolds, see p. 36.
22. John Russell Reynolds, ed., *A System of Medicine* (London: Macmillan, 1868). Bastian contributed in volume 2 the descriptions of pathology and morbid anatomy for pp. 413–429, "Congestion of the Brain"; pp. 430–433, "Cerebritis"; pp. 434–477, "Softening of the Brain"; and pp. 478–503, "Adventitious Products in the Brain." He had also published a detailed study "On the Pathology of Tubercular Meningitis," *Edinburgh Medical Journal* 12: 875–901 (1866–1867).
23. T. Buzzard to W. Gowers, 6 Dec. 1905, Royal College of Physicians, auto-

graph letters, Buzzard collection #23. On Buzzard, see Holmes, *National Hospital*, p. 37.

24. On Bastian's ongoing neurological research, see "On the Various Forms of Loss of Speech in Cerebral Disease," *BFMCR* 43: 209–236, 470–492 (1869); and "Consciousness," *Journal of Mental Science* 15: 501–523 (1869–70). His work is discussed in Anne Harrington, *Medicine, Mind and the Double Brain* (Princeton: Princeton University Press, 1987), pp. 49–50, 218–219, 225. Also in Rang, *Life and Work*, sect. 1. A brief bio in Webb Haymaker, ed., *Founders of Neurology* (Springfield, Ill.: Charles Thomas, 1953), pp. 241–244, describes him as, with Hughlings Jackson, one of the foremost founders of British neurology.
25. Bastian, *On the 'Muscular Sense' and On the Physiology of Thinking* (London: H. K. Lewis, 1869).
26. See Bastian, *Beginnings of Life*, v. 1 (London: Appleton & Co., 1872), pp. 35–47. On psychophysical parallelism as a constitutive element in "scientific naturalism," see Frank Turner, *Between Science and Religion* (New Haven: Yale University Press, 1974), p. 15.
27. W. Sharpey to Bastian, 24 Apr. 1868, Wellcome Institute London, autograph letters collection. This research resulted in a note on "Passage of Red Blood Corpuscles Through the Walls of the Capillaries in Mechanical Congestion" in the *BMJ* 1: 425–426 (1868).
28. Bastian, *MMJ* 1 (Feb. 1869), *Scientific Opinion* 1: 330–331 (1869).
29. Arthur E. Durham, "President's Address, Delivered at the Annual Meeting, July 24th, 1868," *Journal of the Quekett Microscopical Club* 1: 95–110 (1868–69), quotation on p. 110.
30. On Macmillan's crucial role, see Roy MacLeod, "Macmillan and the Young Guard," *Nature* 224: 435–461 (1969). On Youmans, see a collection of essays edited by him arguing the importance of science education, *Modern Culture: Its True Aims and Requirements* (London: Macmillan, 1867).
31. See Ethel Fiske, *The Letters of John Fiske* (New York: Macmillan, 1940), pp. 109–111.
32. Francis E. Abbot, "Review of *The Principles of Biology* by Herbert Spencer," *North American Review* 107: 377–422 (1868).
33. Spencer to Youmans, Mar. 1869, *LLHS*, p. 190. For more on Edward Frankland and his role in the X Club, see Colin Russell, *Edward Frankland: Chemistry, Controversy and Conspiracy in Victorian England* (Cambridge: Cambridge University Press, 1996).
34. Spencer's reply "On Alleged 'Spontaneous Generation' and on the Hypothesis of Physiological Units," written 5 Dec. 1868, was eventually published as an appendix to the next (1870) edition of *Principles of Biology*, v. 1, pp. 479–492.
35. Youmans to Spencer, Mar. 1869, *LLHS*, pp. 190–191.

36. Spencer, "On Alleged 'Spontaneous Generation,'" p. 479.
37. [Bastian], "The Doctrine of the Correlation of the Vital and Physical Forces," *BMJ* 1: 50–51; "Vital Functions and Vital Structures," *BMJ* 1: 288–289; and the seven-part "The Origin of Life," *BMJ* 1: 312–313, 569–570 and *BMJ* 2: 157–158, 214–215, 270–272, 473–474, 665.
38. Alison Adam, "Spontaneous Generation in the 1870s: Victorian Scientific Naturalism and its Relation to Medicine," Ph.D. diss., Sheffield Hallam University, 1988, pp. 200–202.
39. John Browning, "On the Correlation between Microscopic Physiology and Microscopic Physics," printed by Lawson in both *MMJ* 2: 15–21 and *Scientific Opinion* 2: 347–348 (1869).
40. A. E. Durham, "Presidential Address of July 23rd, 1869," *Journal of the Quekett Microscopical Club* 1: 240–248 (1868–69), esp. pp. 244–247.
41. The two men had just served together as joint Secretaries to the Medical Section at the annual meeting of the British Medical Association in Leeds in August 1869. See Humphry D. Rolleston, *The Right Honourable Sir Thomas Clifford Allbutt: A Memoir* (London: Macmillan, 1929), p. 43. Indeed, the two were such close friends that Bastian later said he recognized so much of Allbutt in the fictional Dr. Lydgate of *Middlemarch,* that he was able while calling on G. H. Lewes and George Eliot one day in 1872, to get Eliot to confess "that Dr. Allbutt's early career at Leeds had given her suggestions" for the figure of Lydgate. See H. Cushing, *Life of Osler* (London: Oxford University Press, 1940), v. 1, p. 463n. Bastian's 1869 pamphlet *On the 'Muscular Sense' and On the Physiology of Thinking* was in Lewes and Eliot's library (see Sally Shuttleworth, *George Eliot and Nineteenth Century Science,* Cambridge: Cambridge University Press, 1984, p. 236).
42. Allbutt to Bastian, 10 Dec. 1869, Wellcome Institute London, autograph letters collection.
43. Lawson, *Scientific Opinion* 1: 483 (1869).
44. Stirling to Bastian, 6 Nov. 1869, Wellcome Institute London, autograph letters collection. On Masson, see Frank Turner, "Lucretius Among the Victorians," in *Contesting Cultural Authority* (Cambridge: Cambridge University Press, 1993).
45. H. C. Bastian, "Protoplasm," *Nature* 1: 424–426 (24 Feb. 1870), quotation on p. 426. After Bastian's scathing review, it is probable that Stirling considered this promising young talent to be headed down a path that would waste his energies.
46. Wyman to Wilder, 25 Nov. 1869, Wilder papers, Box 1, Cornell University Archives: "while I do not believe spontaneous generation proved, I by no means consider it disproved."
47. Wyman to Wilder, 16 Oct. 1870, Wilder papers, Box 1, Cornell University Archives.

48. Wyman to Bastian, 21 Oct. 1870, APS Library, 509 L56.42, misfiled under "Bartram." I am indebted to Toby Appel for this reference and the preceding two. Charles Darwin, even during his more skeptical moments about spontaneous generation, still felt that Wyman's experiments had to be explained before one could dismiss the subject, whatever one thought of Bastian's experimental skill.
49. Thomas R. R. Stebbing, "Note on the Hypothesis of Spontaneous Generation," pp. 126–146 in *Essays on Darwinism* (London: Longmans, Green, 1871), pp. 127–128.
50. Wallace, "Review of *The Beginnings of Life*," *Nature* 6: 284–287, 299–303 (8 Aug. and 15 Aug. 1872), esp. pp. 302–303; Wallace to Darwin, 4 Aug. 1872, in *ARWLR*, v. 1, pp. 224–225 (*DCC* #8450).
51. Wallace, "Review of *The Beginnings of Life*," p. 303.
52. Darwin to Wallace, 28 Aug. 1872, in *ARWLR*, v. 1, pp. 225–226 (*DCC* #8488); an expurgated version is in *LLCD*, v. 3, pp. 168–169. Darwin expressed himself similarly to Haeckel on 2 Sept. 1872 (*DCC* #8506; Darwin papers, APS Library):

    Our English Dr. Bastian has lately published a book on so-called Spontaneous Generation, which has perplexed me greatly. He has collected all the observations made by various naturalists, some of them good observers, on the protoplasm within the cells of dying plants and animals becoming converted into living organisms. He has also made many experiments with boiled infusions in closed flasks; but I believe he is not a very careful observer. Nevertheless, the general argument in favor of living forms being now produced under favorable conditions seems to me strong; but I can form no final conclusions.

    See also Darwin's copious notes in his copy of the book, in M. A. DiGregorio, ed., *Charles Darwin's Marginalia* (New York: Garland, 1990), pp. 34–35.
53. Wallace to Darwin, 31 Aug. 1872, *ARWLR*, p. 226 (*DCC* #8498).
54. See Macmillan to Youmans, 1 Feb. 1872, Macmillan papers, Ms. 55392 (1), p. 164, British Library.

    When you were in England you expressed a strong wish that you could have the sale of Dr. Bastian's book "Beginnings of Life" for America, secured to Messrs. Appleton. Dr. Bastian has spoken to me about it also somewhat urgently, and as I have a desire to gratify him & also to show my good will to you, I have consented to make you the offer of the edition which we are printing in one volume for the American market. It consists of 750 copies and will be printed in one volume, the English one being in two. The volume will be over 600 pages and will have over 80 very carefully executed illustrations. They have cost us nearly £150.

> What I propose is that we should send the whole of this edition over to Messrs. Appleton, charging 3s/9d a copy . . ., leaving them to pay to Dr. Bastian their usual royalty. They will thus secure the American market and if a new edition is shortly called for we will supply them with *clichés* of the cuts and early sheets of our reprint & they can then print & arrange with Dr. Bastian himself . . . The book will be ready in about six weeks or two months.

55. [Youmans], "Spontaneous Generation," *Popular Science Monthly* 2: 83–93 (Nov. 1872), esp. pp. 91–93.
56. Bastian, "Evolution and the Origin of Life," *Popular Science Monthly* 4: 713–728 (1874).
57. Youmans to Bastian, 16 Apr. 1874, Wellcome Institute London, autograph letters collection.
58. Youmans, "Sketch of Dr. H. C. Bastian," *Popular Science Monthly* 8: 108–110 (Nov. 1875).
59. Personal communication 15 Oct. 1994 & 5 Feb. 1996 from British historian Michael Collie. See also Bastian to Owen, 28 Oct. 1876, Owen papers, BMNH 2: 324–326. The book *The Brain as an Organ of Mind* was published in 1880. Collie notes "despite the delay, the publisher's enthusiasm for the book was demonstrated by what for the series as a whole was an unusually high initial print run of 2500 copies, instead of 1250."
60. John Fiske, *Edward Livingston Youmans: Interpreter of Science* (New York, 1885), p. 320.
61. Holmes, *National Hospital,* p. 39.
62. James Murie, "On the Development of Vegetable Organisms Within the Thorax of Living Birds," *MMJ* 7: 149–162 (1872), p. 150.
63. Bastian to Murie, 4 Apr. 1872, Murie Letters, Linnean Society Library, Burlington House, London.
64. *Annual Register* 114 (1872), p. 368, italics mine, quoted in James Friday, "A Microscopic Incident in a Monumental Struggle," *British Journal for the History of Science* 7: 61–71 (1974).
65. *Munk's Roll of the Royal College of Physicians* (London, 1870), v. 1, pp. 174–175.
66. William Osler, "The Medical Clinic: a Retrospect and a Forecast," *BMJ* 1: 10–16 (3 Jan. 1914), p. 10; also in H. Cushing, *Life of Osler* (London: Oxford University Press, 1940), v. 1, p. 88, see also p. 100.
67. Ghetal Burdon Sanderson, *Sir John Burdon Sanderson: A Memoir* (Oxford: Clarendon Press, 1911), pp. 23–25, 51.
68. It was published shortly thereafter. William Osler, "An Account of Certain Organisms Occurring in the Liquor Sanguinis," *PRSL* 22: 391–398 (1874); also in *MMJ* 12: 141–148 (1 Sept. 1874).

69. Jabez Hogg, *The Microscope: Its History, Construction, and Application* (London: Routledge and Sons) had gone through six editions by 1867. On his support of Bastian, see his participation in the 1875 Pathological Society debate on the germ theory of disease, *QJMS* 15: 327–328, and Bastian to Hogg, 12 May 1875, Wellcome Institute London, autograph letters collection.
70. William Bulloch, *History of Bacteriology* (London: Oxford University Press, 1938), p. 184; Lewis to Bastian, 15 July 1873, Wellcome Institute London, autograph letters collection.
71. Sharpey to Bastian, 16 Nov. 1871, Wellcome Institute London, autograph letters collection. On Bastian's request for a research grant, see Bastian to Sharpey, 17 Oct. and 2 Nov. 1870, RS MC.9.129 and 133.
72. Bastian, "Bacteria," "Germs of Disease," "Aphasia" in *Quain's Dictionary.* Bastian's articles on these topics continued to be included even up to the 1890 edition of the *Dictionary.*
73. Macmillan to Bastian, 7 Jan. 1870, Macmillan letterbook, Ms 55390, British Library.
74. Bastian, "Facts and Reasonings Concerning the Heterogeneous Evolution of Living Things," *Nature* 2: 170n (30 June 1870).
75. Macmillan to Bastian, 3 Jan. 1870; *Nature* 1:253 (title page of 6 Jan. issue), (20 Jan., p. 301), (17 Feb., p. 398), (3 Mar., p. 450), (10 Mar., p. 474), (28 Apr., p. 668).
76. *PRSL* 20: 239–264 (1872).
77. This paper was read 30 Jan., voted yes 22 Feb. for publication, and appeared in *PRSL* 21: 129–131 (1873). It was around the time of receipt by the Society of this paper that Burdon Sanderson a few days later contacted Bastian to arrange to work with him in Bastian's lab at UCL to replicate some of the experiments. This led him by 1 Jan. to begin writing a paper for *Nature* that confirmed many of Bastian's results.
78. Bastian's paper, read 20 Mar. and voted for publication the same evening, appeared in *PRSL* 21: 224–232; his paper read 15 May, voted for publication the same evening, appeared in *PRSL* 21: 325–338. It should be noted, however, that publication in the *Proceedings* was not nearly as restricted and therefore prestigious as publication in the *Philosophical Transactions,* which required favorable reports by two FRS referees chosen confidentially by the Council. The reports were also kept confidentially in Royal Society Archives. See chapter 7 for more on this subject. Recall that Bastian's first paper to the Royal Society, on Nematoid worms, was voted to appear in the *Philosophical Transactions* after glowing reviews from Busk and Huxley.

4 Initial Confrontation with the X Club

1. See *BMJ* 2: 632 (10 Dec. 1870) for a sample of the conflict seen by the actors themselves as a professional turf battle. Interestingly, Beale, though opposed to Bennett, Bastian, and spontaneous generation, was even more vituperatively opposed to Tyndall's attempt to make proclamations about disease and microscopy.
2. James Moore, "Deconstructing Darwinism: The Politics of Evolution in the 1860s," *Journal of the History of Biology* 24: 353–408 (1991), pp. 386–387.
3. George H. Lewes, "Mr. Darwin's Hypothesis," *Fortnightly Review* n.s. 3: 353–373 (1 Apr. 1868), quotation on p. 357n.
4. Thomas H. Huxley, "Yeast," in *Contemporary Review*, Dec. 1871, reprinted as pp. 113–140 in *Discourses Biological and Geological* (London: Appleton & Co., 1925). The response to Stirling is a few paragraphs on pp. 136–139 basically dismissing Stirling as an armchair critic, not to be taken seriously because he would never think to look into a microscope to confirm for himself the claims of those he cites.
5. T. H. Huxley, "Biogenesis and Abiogenesis," pp. 232–274 in *Discourses Biological and Geological.*
6. By 21 Oct. 1870, Wyman (letter to Bastian) seems to have also come around to this "at least in the distant past" position in support of evolution. However, he still seems to have some sympathy for spontaneous generation or hope for experimental proof. The position of Friedrich Engels at that time was quite similar. He had sympathy for the possibility of an evolutionary process of "structureless Monera . . . from structureless living protein" directly derived from his reading of Haeckel and Darwin. See Engels, *Dialectics of Nature,* ed. J. B. S. Haldane (New York: International Pub., 1940), pp. 188–189.
7. These criticisms were published soon thereafter in Huxley's article "On the Relations of Penicillium, Torula, and Bacterium," *QJMS* 10: 355–362 (1870), pp. 359–362.
8. Wallich to Darwin, 25 Mar. 1882, *DCC* #13739, Darwin papers, APS Library, Philadelphia. Darwin seems to have agreed, for Wallich went ahead with this point in his talk to the Victoria Institute (the Philosophical Society of Great Britain) on 17 Apr. 1882, published in *Transactions of the Victoria Institute* 16: 344 (1882).
9. Dale A. Johnson, "Popular Apologetics in Late Victorian England: The Work of the Christian Evidence Society," *Journal of Religious History* 11: 558–577 (1981), pp. 558, 559.
10. Ibid., p. 559.

11. See the detailed discussion of Dallinger's work and the endnotes to chapter 5 for the citations to these influential articles.
12. Anon., "The Atmospheric Germ Theory," *Nature* 1: 351 (3 Feb. 1870).
13. See Joseph Lister, for instance. M. Pelling (*Cholera, Fever, and English Medicine, 1825–1865* [Oxford: Oxford University Press, 1979], pp. 134–136) has noted that, in the traditional descriptions of responses of the medical community to the debate between Pasteur and Liebig over theories of fermentation and disease, Pasteur's account of the total incompatibility of his work on "germs" with Liebig's has been accepted uncritically. She shows that, especially in the British medical community, the two men's work was not seen as necessarily incompatible, and many intermediate positions were taken in British disease theory through the late 1860s.
14. In his first public statement on spontaneous generation in his 7 Apr. 1870 letter to the *Times,* Tyndall's argument sounds as if it was Lister's dramatic results that spurred him to form an opinion on spontaneous generation.
15. Huxley to Dohrn, 4 July 1868, Huxley papers 13: 166–167. The expression comes from the play *Henry V,* V-I: 10, "look you now, *of no merits,* he is come to me, and prings me pread and salt yesterday, look you, and bid me *eat my leek.*" My thanks to John Dettloff for help in locating this passage.
16. Huxley to Foster, 11 Aug. 1875, Royal College of Physicians, autograph letters collection.
17. *LLTHH,* v. 2: 5.
18. Foster to Huxley, 14 Sept. 1879, Huxley papers 4: 214–215.
19. Walter White, *The Journals of Walter White, Assistant Secretary of the Royal Society* (London: Chapman & Hall, 1898), p. 271.
20. Huxley to Dohrn, 30 Apr. 1870, Huxley papers 13: 174.
21. Tyndall to Bastian, 25 Jan. 1870; letters between Macmillan and Bastian, Macmillan letterbook, Ms. 55390, British Library.
22. X Club notebook, Tyndall papers 4/B8, on meeting of 3 Feb. 1870. Anon., "The Atmospheric Germ Theory," *Nature* 1 (3 Feb. 1870): 351. Bastian's review of Stirling, defending Huxley on "Protoplasm" and arguing for a connection between Huxley's "Physical Basis of Life" and spontaneous generation, also appeared in *Nature,* on 24 Feb., pp. 424–426.
23. *LLTHH* v. 1: 355. This version by Huxley's son claims that Huxley's own experimental work had already been underway on this subject for two years, because of "the preparation of the course of Elementary Biology." This would have been stimulated initially by the claims of Hallier and by Huxley's thinking around *Bathybius* and "The Physical Basis of Life" in late 1868. Many parallels are obvious if we compare the norms of behavior Huxley was trying to uphold with those discussed by Steven Shapin and Simon Schaffer in *Leviathan and the Air Pump* (Princeton: Princeton Uni-

versity Press, 1985). Both "gentlemanly conduct" and accumulating credibility through involvement of respected "witnesses" are relevant.

24. Macmillan to Beale, 21 Apr. 1870, Macmillan letterbook, Ms. 55390 (2): 891, British Library.
25. Bastian to Huxley, 2 May 1870, Huxley papers 10: 238.
26. Ibid., Huxley's note on margin of Bastian's letter, indicating the content of his reply.
27. Bastian to Huxley, 12 May 1870, Huxley papers 10: 239.
28. Ibid.
29. Huxley to Dohrn, 30 Apr. 1870, Huxley papers 13: 174 (*LLTHH* 1: 357–358).
30. Hooker to Darwin, 7 July 1870, *DCC* #7267, Darwin papers, APS Library.
31. Darwin to Hooker, 12 July 1870, *DCC* #7273, *MLCD* v. 1: 321–322.
32. Huxley to Hooker, 10 Aug. 1870, Huxley papers 2: 163.
33. Huxley to Tyndall, 15 June 1870, Huxley papers 8: 80.
34. Anon., "Obituary: Robert Brown," *Athenaeum*, 19 June 1858, p. 786.
35. John B. Dancer, "Remarks on Molecular Activity as Shown Under the Microscope," *Proceedings of the Literary and Philosophical Society of Manchester* 7: 162–164 (1868), p. 164.
36. Christian Wiener, "Erklärung des Atomistischen Wesens des Tropfbarflussigen Körperzustandes, und Bestatigung Desselben durch die Sogenannten Molecularbewegung," *Annalen der Physik und Chemie* 118: 79–94 (1863), p. 85. Thanks to Uwe Hollerbach for assistance with translation.
37. Dancer, "Remarks on Molecular Activity," pp. 162–163.
38. Steven Brush, "Brownian Movement from Brown to Perrin," *Archives of History of the Exact Sciences* 5: 1–36 (1968), p. 6.
39. Following Huxley's presidential address, which opened the meeting, a session on spontaneous generation in the biological subsection on Friday 16 Sept. 1870 was marked by "the attention with which the crowded audience listened to a two-hours' discussion." Neither Bastian nor Huxley was present at this session, which included Dr. Child in defense of Abiogenesis and "Mr. Samuelson, . . . Dr. Hooker, Mr. Bentham, and other distinguished naturalists, entirely on the orthodox side of Biogenesis." It was in a session on the following Wednesday, 21 Sept. that Huxley gave his detailed critique of Bastian on Brownian movement, and at which Bastian rose to defend himself during the discussion. See "The British Association," *Nature* 2: 408–410 (22 Sept. 1870), and *Lancet* 2: 514 (8 Oct. 1870).
40. Huxley, "Relations of Penicillium" pp. 358–359.
41. Henry C. Bastian, *The Beginnings of Life*, v. 1 (London: Appleton & Co., 1872), p. 319. Bastian seems to have gotten the date wrong on which Huxley's paper was given, first publicly raising this issue.

42. Ibid.
43. See Henry C. Bastian, *The Modes of Origin of Lowest Organisms* (London: Appleton & Co., 1871), pp. xi–xii. For a recent critique, similar to Bastian's, of Huxley's way of using "biogenesis," see Dean Kenyon, "On Terminology in Origin of Life Studies," *Origins of Life* 6: 447–448 (1975). My historical analysis puts the controversy between Kenyon and Carl Sagan in an interesting new light.
44. Michael Foster, *A Text Book of Physiology* (London: Macmillan & Co., 1877), p. 2n.
45. See the anonymous review of *Beginnings of Life* in the medical journal *The Practitioner* 9: 291–294 (1872).
46. Note that Huxley's distinction of "Xenogenesis" (for what Bastian and most others called heterogenesis) did not seem to catch on. I would suggest that this is because the simpler opposed duality of biogenesis/abiogenesis lent itself better to the condensed version of the story required for the limited space in a biology text. As late as 1910, Huxley's trainees were trying to extend their successful hijacking of "biogenesis." In the closing sentence of his article on "Abiogenesis" for the eleventh edition of the *Encyclopedia Britannica* (v. 1, pp. 64–65), P. Chalmers Mitchell tried to extend the credit to Huxley for "archebiosis" as well.
47. Anon., "Bastian on Lowest Organisms," *BFMCR* 49: 450 (1872).
48. Huxley, "Biogenesis and Abiogenesis," pp. 258–260.
49. Huxley to Hooker, 11 Aug. 1871, Hooker papers, tss. pp. 101–102, RBG Kew.
50. Hooker to Darwin, 5 Aug. 1871, *DCC* #7896, *LLJDH* 2: 126–127. To Huxley, Hooker was even more blunt. He loved the "cockshy" expression, and continued: "Thomson is of the Scotch, Scotchy. And what a scratching of Tait: I suppose they lay their heads together (Tait à Tait) and communicated living matter as well as ideas." Hooker to Huxley, 19 Aug. 1871, Hooker papers, Kew, tss. pp. 109–110.
51. Hooker to Darwin, 5 Aug. 1871, *DCC* #7896. Tyndall, too, was amused yet irked by Thomson's idea:

> It amused and interested me. Nobody but a man of Thomson's brilliancy of intellect would have alighted upon that idea of peopling the earth by organic seed derived from meteorites . . . Why should not the seeds be developed in the earth as well as in the meteorites, and if the fiat of a creator be necessary why should it not apply to the earth as well as to meteoric matter? Perhaps the report I have seen is an incomplete or an incorrect one; but if it be correct I cannot help wishing that Thomson had continued to muse over it in private instead of embodying it in his address. What an enormous influence early education exercises on some men [re-

> ferring to Thomson's religious upbringing]! I dare say it was such an influence that caused Thomson to take refuge in this hypothesis rather than accept the notion that life came direct from what has been incorrectly termed inorganic matter. I dare say amid his thousand other avocations he has had but little time to dwell in the presence of this problem; if he had I hardly think he would have tried to dispose of it by a mere brilliant *conceivability.*

Tyndall to Bence Jones, 13 Aug. 1871, Tyndall papers, tss. pp. 805. For more on the conflicts Tyndall and the X Club had at this time with Thomson, Tait, and Maxwell, see Crosbie Smith, *The Science of Energy* (London: Athlone, 1998), chapter 9.

52. See Steven Dick, *The Biological Universe* (Cambridge: Cambridge University Press, 1996); also Dick, *Life on Other Worlds* (Cambridge: Cambridge University Press, 1998).
53. Darwin to Hooker, 1 Feb. 1871, *DCC* #7471, Darwin papers, APS Library, Philadelphia. This quotation was first brought to light in the twentieth-century origin of life research community in an article by Garrett Hardin, "Darwin and the Heterotroph Hypothesis," *Scientific Monthly* 70: 178–179.
54. Bastian, *Nature* 2: 411–413, 431–434, 473, 492 (1870).
55. Huxley, "Dr. Bastian and Spontaneous Generation," *Nature* 2: 473 (13 Oct. 1870).
56. Huxley to Lockyer, 8 Oct. 1870, Huxley papers 21: 252–253.
57. Peter Bowler, "Darwinism in Britain and America," in D. Kohn, ed., *The Darwinian Heritage* (Princeton: Princeton University Press, 1985), p. 668.
58. William Thistleton-Dyer, "On Spontaneous Generation and Evolution," *QJMS* 10: 333–354 (1870), quotation on p. 335.
59. Darwin to Dohrn, 3 Feb. 1872, *The Naples Zoological Station at the Time of Anton Dohrn,* ed. Christiane Groeben (Woods Hole, Mass., 1980), pp. 84–85 (*DCC* #8199).
60. Dohrn to Darwin, 21 Aug. 1872, *Charles Darwin–Anton Dohrn Correspondence,* ed. Christiane Groeben (Naples: Macchiaroli, 1982), pp. 40–41 (*DCC* #8481).
61. See Fiske's letters to his wife of Dec. 1873 and Jan. 1874 in E. Fiske, ed., *The Letters of John Fiske* (New York: Macmillan, 1940), pp. 270–301.
62. J. Fiske, *Outlines of Cosmic Philosophy, Based on the Doctrine of Evolution* (London: Macmillan, 1875), v. 1, p. 129.
63. John Fiske, *The Unseen World* (Boston: Houghton Mifflin, 1876), p. 49.
64. William H. Brock, "Patronage and Publishing: Journals of Microscopy, 1839–1989," *Journal of Microscopy* 155: 249–266 (1989), p. 253. Both of the latter two journals were published and financially supported by the naturalist publisher Robert Hardwicke until his death in 1875.

65. Ruth Barton, "Just Before *Nature:* the Purposes of Science and the Purposes of Popularization in Some English Popular Science Journals of the 1860s," *Annals of Science* 55: 1–33 (1998).
66. Lawson, "Spontaneous Generation," *Popular Science Review* 10: 416–417 (Oct. 1871).
67. [Lawson], "On Spontaneous Generation," *Popular Science Review* 14: 184–185 (Apr. 1875).
68. Ibid., p. 184.
69. Marie Boas Hall, *All Scientists Now: The Royal Society in the Nineteenth Century* (Cambridge: Cambridge University Press, 1984), p. 136.
70. Bastian to Huxley, 31 Jan. 1873, MC.9.484, RS Archives. Huxley's handwritten note at the top of Bastian's letter says "Pleased to insert Dr. Bastian's note as below, dating it (Jan. 31 1873)."
71. Secretaries of the Royal Society exercised enormous powers over many such matters, including the awarding of research grants. See Boas Hall, *All Scientists Now*, pp. 135–136, and Roy MacLeod, "The Royal Society and the Government Grant: Notes on the Administration of Scientific Research, 1849–1914," *Historical Journal* 14: 323–358 (1971), esp. pp. 339–340, 343, 348–350.
72. Bastian to Huxley, 2 May 1873, MC.9.543, RS Archives.
73. Huxley to Bastian, 5 May 1873, MC.9.545, RS Archives.
74. Bastian to Huxley, 11 May 1873, MC.9.551, RS Archives.
75. Bastian to Huxley, 14 May 1873, MC.9.553, RS Archives.
76. Ibid.
77. Bastian to Huxley, 18 May 1873, MC.9.555, RS Archives.
78. Ibid.
79. Bastian to Huxley, 19 May 1873, MC.9.557, RS Archives.
80. Ruth Barton, "'An Influential Set of Chaps': The X Club and Royal Society Politics 1864–85," *British Journal for the History of Science* 23: 53–81 (1990).
81. MacLeod, "Royal Society," p. 343. Had Huxley been secretary in the fall of 1870 when Bastian applied with Frankland for a research grant for archebiosis experiments, it is worth wondering whether the grant would have been awarded. As noted previously, Sharpey was still the secretary for biological sciences at that time, and he still considered the work important enough to award the grant (Bastian to Sharpey, 17 Oct. and 2 Nov. 1870, RS, MC.9.129, and 9.133).
82. Huxley to Bastian, 19 June 1873, MC.9.587, Royal Society Archives.
83. C. C. Pode and E. Ray Lankester, "Experiments on the Development of Bacteria in Organic Infusions," *PRSL* 21 (1872–3): 349–358. Also appeared in *MMJ* 10: 118–128 (1 Sept. 1873).

84. E. Ray Lankester, "Ernst Haeckel on the Mechanical Theory of Life and on Spontaneous Generation," *Nature* 3: 354–356 (2 Mar. 1871).
85. Dohrn to Huxley, 19 Apr. 1872, and Huxley to Dohrn, 5 June 1872, Huxley papers 13: 212–213, 217–219. Huxley mentioned helping Lankester get a fellowship as an instructor at Oxford, though noting that Lankester would have to relinquish a love affair with a woman he had met in Naples if he accepted the fellowship (8: 222–223). For more on the explosive, "loose cannon" temperament that Huxley had foreseen, see Joe Lester, *E. Ray Lankester and the Making of Modern British Biology*, ed. Peter Bowler, *British Society for the History of Science Monographs* 9 (Oxford: BSHS, 1995).
86. Lankester, "Review of Bastian's *Beginnings of Life*," *QJMS* 13: 59–74 (1 Jan. 1873), pp. 70–71.
87. See Lister to Dyer, 10 Sept. 1873, Dyer papers, RBG Kew; Joseph Lister, "A Further Contribution to the Natural History of Bacteria and the Germ Theory of Fermentative Changes," *QJMS* 13: 380–408 (1873).
88. Lester, *E. Ray Lankester.*
89. Lankester to Dohrn, 24 Apr. 1875, in *Naples Zoological Station*, pp. 88–89.
90. Henry Acland to Edward Sieveking, 5 June 1873, RCP autograph letters collection.
91. Lankester to Huxley, 18 Dec. 1872, Huxley papers 21: 39–44.
92. Bastian, "Spontaneous Generation," *Nature* 9: 483 (23 Apr. 1874).

5 Colloids, Pleomorphic Theories, and Cell Theories

1. John Crellin, "The Dawn of the Germ Theory: Particles, Infection and Biology," in F. N. L. Poynter, ed., *Medicine and Science in the 1860s* (London: Wellcome Institute, 1968), p. 58.
2. William Coleman, "Cell, Nucleus and Inheritance: An Historical Study," *Proceedings of the American Philosophical Society* 109: 124–158 (1965), pp. 125, 128.
3. See the obituary on Graham by Alexander Williamson in *Nature* 1: 20–22 (4 Nov. 1869); also H. Hale Bellot, *University College London* (London: London University Press, 1929), pp. 127–130; Gerald L. Geison, *Michael Foster and the Cambridge School of Physiology* (Princeton: Princeton University Press, 1978), p. 39; Henry Bence Jones, *Lectures on Some of the Applications of Chemistry and Mechanics to Pathology and Therapeutics* (London: John Churchill, 1867), p. 13; R. D. C. Black and R. Könekamp, eds., *Papers and Correspondence of William Stanley Jevons*, v. 1 (London: Macmillan, 1972), pp. 14–15, 69–70, 101, 114.
4. Black and Könekamp, eds., *Papers of Jevons*, p. 14.
5. Margaret Pelling, *Cholera, Fever, and English Medicine, 1825–1865* (Oxford:

Oxford University Press, 1979), p. 142. William Farr developed his idea of "zymotic particles" or "zymads" based directly on Graham's suggestion and Liebig's theory (see Pelling, p. 107).

6. Thomas Graham, "Liquid Diffusion Applied to Analysis," *PTRSL* 151: 183–224 (1861).
7. Ibid., p. 183, italics in original.
8. Ibid., p. 184.
9. By Dec. 1871 Huxley was jokingly describing his near-nervous breakdown by referring to himself as "a poor devil whose brains and body are in a colloide [sic] state." Huxley to Tyndall 22 Dec. 1871, Huxley papers.
10. Benjamin W. Richardson, "The Germ Theory of Disease," *BMJ* 2: 467 (1870). See also his book *Diseases of Modern Life* (London: Macmillan, 1876), pp. 81–92.
11. Henry Maudsley, "The Theory of Vitality," *BFMCR* 32: 400–423 (1863). On how well-regarded Maudsley's article was, see *Practitioner* 9 (1872): 291–292. For details on another theory from this period about the structure of matter that was intimately linked to support for evolution and spontaneous generation, see J. S. Wilkie, "Nägeli's Work on the Fine Structure of Living Matter," *Annals of Science* 16: 11–41 (1960).
12. Maudsley, "Theory of Vitality," p. 409.
13. Ibid., p. 419.
14. Ibid., p. 420.
15. Ibid., p. 421.
16. Margaret Schabas, *A World Ruled By Number: William Stanley Jevons and the Rise of Mathematical Economics* (Princeton: Princeton University Press, 1990).
17. William S. Jevons, "On the So-Called Molecular Movements of Microscopic Particles," *Scientific Opinion* 2: 155 (1869); full text in *Proceedings of the Manchester Literary and Philosophical Society* 9 (ser. 3): 78–84 (1870). See also Dancer's paper on Brownian movement in the same *Proceedings* v. 7: 162–164 (1868).
18. Jevons, "On the So-Called Molecular Movements," p. 155.
19. Benjamin Lowne, "On So-Called Spontaneous Generation," *Journal of the Quekett Microscopical Club* 2: 133–140 (1870–71).
20. John Farley claims, "Although Graham himself had drawn attention to the biological implications of colloid behavior, it was not until the 1890s that the colloid nature of . . . protoplasm was first seriously studied." (See *The Spontaneous Generation Controversy from Descartes to Oparin* [Baltimore: Johns Hopkins University Press, 1977], p. 157.) This section has shown that between 1861 and 1877 or so, Farley's statement is misleading.
21. "Obsolete" is a relative term historically. This was proven when Liverpool

biochemist Ben Moore revived Bastian's theories and interpreted his language about spontaneous generation using the new colloid jargon of the 1910s and 1920s. Moore even attempted to bring Graham's "energia" back from the obscurity to which it had been relegated. See Farley, *Spontaneous Generation*, pp. 155–165 and Moore, *The Origin and Nature of Life* (London: Williams and Norgate, 1913) reprinted in 1921 (page refs. are to the later edition), pp. 157, 225–226. Moore was trying to put a new name, "biotic energy," on what he saw as the specific life energy.

22. On Mayer, see Ken Caneva, *Robert Mayer and the Conservation of Energy* (Princeton: Princeton University Press, 1993). For more on Robert Grove's "correlation of forces" and its relationship to his larger philosophical and practical reform commitments, see Iwan Rhys Morus, "Correlation and Control: William Robert Grove and the Construction of a New Philosophy of Scientific Reform," *Studies in History and Philosophy of Science* 22: 589–621 (1991). On the role of Mayer, Helmholtz, and Grove in developing the conservation of energy and "correlation of forces" doctrines, see Thomas Kuhn, "Energy Conservation as an Example of Simultaneous Discovery," in *The Essential Tension* (Chicago: University of Chicago Press, 1974), pp. 66–104.
23. Henry Bence Jones, *Lectures on Applications of Chemistry and Mechanics to Pathology and Therapeutics* (London: John Churchill, 1867); Bence Jones, *Croonian Lectures on Matter and Force* (London: Churchill, 1868).
24. On the atom debates in the Chemical Society at this time, see Henry Lawson, "The Battle of the Atoms," *Scientific Opinion* 2: 485 (1869); or a more recent secondary source, David Knight, *Atoms and Elements* (London: Hutchinson, 1967), pp. 51–52, 104–126. On the separability of force and matter, see Bence Jones, *Croonian Lectures*, p. 32, and on the aether question, pp. 20–26, which shows Maxwell's opposition to Grove's general schema. Bastian and most British spontaneous generation advocates were heavily committed to Grove. This may have been yet another reason why Maxwell was violently opposed to spontaneous generation and pangenesis, in addition to his deep antagonism to Darwinism in general. For more on Maxwell, Thomson, and Tait's hostility toward evolution, Grove, the X Club, and especially Tyndall, see Crosbie Smith, *The Science of Energy* (London: Athlone, 1998), chapter 9, especially pp. 176–177.
25. Bence Jones, *Croonian Lectures*, p. 32. The connection was widely known since at least the famous 1850 paper by William B. Carpenter to the Royal Society, "On the Mutual Relations of the Vital and Physical Forces," *PTRSL* 140: 727–757 (1850). See also Carpenter's review of the new cell theory in 1840, "Schwann and Schleiden on the Identical Structure of Plants and Animals," *British and Foreign Medical Review* 9: 495–528. In that review we

see Carpenter's interest (early in Britain) in Schwann's "organic molecules" in histology.

26. Frank M. Turner, *Contesting Cultural Authority: Essays in Victorian Intellectual Life* (Cambridge: Cambridge University Press, 1993), p. 146. Carpenter actually became increasingly religious from the late 1840s onward. See Vince M. D. Hall, "The Contributions of the Physiologist, William Benjamin Carpenter (1813–1885), to the Development of the Principles of the Correlation of Forces and the Conservation of Energy," *Medical History* 23: 129–155 (1979), esp. pp. 152–155.
27. First stated by Bastian in print on 16 Jan. 1869 in an unsigned editorial, "The Doctrine of the Correlation of the Vital and Physical Forces," in the *BMJ* 1: 50–51. See also anon., "Review of *The Beginnings of Life*," *Practitioner* 9: 291–294 (1872).
28. Owen, *On the Anatomy of the Vertebrates* (London: Longmans, Green, 1868), v. 3: 819–822; John Tyndall, *Faraday as a Discoverer* (London, 1868); Bence Jones, *Croonian Lectures*, pp. 40–41. See also his *Life and Letters of Michael Faraday* (London: Longmans, Green, 1870). In this biography Bence Jones described a lecture by Faraday shortly after Robert Brown's first controversial paper on active molecules (v. 1, pp. 403–404). Faraday clearly stated in the lecture that Brown had *not* made the claims about the movement of the molecules being vital in nature, as many thought he had. This was a response by Bence Jones to the greatly proliferating spontaneous generation advocacy in 1868–69, referring to "living molecules" in a way that strongly evoked that earlier controversy. He, like Tyndall, wished to emphasize the 1868 view of J. B. Dancer on Brownian movement as a purely physical phenomenon.
29. Bence Jones, *Croonian Lectures*, p. 32.
30. John Browning, "On the Correlation of Microscopic Physiology and Microscopic Physics," *Scientific Opinion* 2: 347–348 (1869) and *MMJ* 2: 15–21 (1869).
31. The introduction to the book was titled "The Nature and Source of the Vital Forces and of Organizable Matter."
32. Frederick A. P. Barnard, "The Germ Theory of Disease and Its Relations to Hygiene," *Public Health Reports and Papers* 1: 70–87 (1873).
33. Ibid., p. 79.
34. See Ruth Barton, "John Tyndall, Pantheist: A Rereading of the Belfast Address," *Osiris* 3(n.s.): 111–134 (1987) for a discussion of Carlylean pantheism and the public's consistent misinterpretation of it as materialism.
35. Barnard, "Germ Theory of Disease," pp. 79–80. It is interesting that this passage was one of the very few excerpted from a bowdlerized version of Barnard's essay published in Gert Brieger, ed., *Medical America in the Nine-*

teenth Century (Baltimore: Johns Hopkins, 1972), pp. 278–292; passage expurgated on p. 286.

36. For a much more extensive discussion of the grappling of Victorian intellectuals, including Wallace, with this dilemma in general, see Frank Turner, *Between Science and Religion* (New Haven: Yale University Press, 1974).
37. See Lowne, "On So-Called Spontaneous Generation," pp. 134–138, about which Darwin responded: "It will be a curious discovery if Mr. Lowne's observation that boiling does not kill certain moulds is found true; but then how on earth is the absence of all living things in Pasteur's experiments to be accounted for?" Darwin to Hooker 1 Feb. 1871, *DCC* #7471, Darwin papers, APS Library, Philadelphia.
38. See Lawson's editorials in *Popular Science Review* 7: 176–180, 233–234, 315 (1868); 8: 71, 77, 166–167, 172–173, 185–186 (1869); *MMJ* 3: 316–317, 320 (1870); and *Scientific Opinion* 1: 122, 253–255, 297–299, 483 (1868–69); 2: 216, 248, 391–392, 426, 569–570, 600–601 (1869); 3: 247–248, 337–338, 481 (1870).
39. Lionel S. Beale, *How to Work with the Microscope,* 3rd ed. (London, 1864), pp. 246–247. It is only fair to point out that Bennett was not the only critic who turned Beale's criticism of Bennett back on Beale himself. James Ross said of "bioplasm," "germinal matter," and "formed matter" that "Dr. Beale's terms are good examples of the 'question begging appellatives.'" See "A Critique of Dr. Beale's Theory of Life," *Practitioner* 6: 266–276 (1871), quotation on p. 269.
40. Tyndall, "Dust and Disease" (1872), in *Essays on the Floating Matter of the Air in Relation to Putrefaction and Infection* (New York: Johnson Reprint, 1966), pp. 31–33.
41. In fact, the passages in Bennett's *Textbook of Physiology* (Philadelphia: Lippincott, 1872) on the subject were almost verbatim extracts from his 1863 *Lancet* series on molecules (pp. 35–46, 48–54, 56–60) and his 1868 paper on heterogenesis (pp. 46–48, 50n, 421–440). He added brief notes citing new work by Montgomery, Béchamp, Bastian, Murie, and Trecul as confirming his position. Furthermore, Gilbert Child (*Nature* 1: 626) and Henry Lawson both lauded Bennett's work in agreement with their own support for spontaneous generation, as did H. C. Bastian in his widely read *Beginnings of Life* (London: Appleton & Co., 1872), v. 1, p. 160; v. 2, pp. 59, 344.
42. Francis Galton, "Experiments in Pangenesis," *PRSL* 19: 393–409 (30 Mar. 1871); also "On Blood Relationship," *PRSL,* 13 June 1872. See also James Clerk Maxwell, "Atom," 9th ed. of *Encyclopedia Britannica* 3: 33–44, esp. pp. 37–38 (1875). E. Ray Lankester, however, continued as before to speak

enthusiastically of hereditary "molecules" in the embryo, just as he continued to be an advocate of *Bathybius* and of pleomorphism in the British context into the late 1870s and beyond.

43. Bennett repeated his critique of Virchow on pp. 101–103 of the *Textbook of Physiology*, and repeated his charge against Beale on pp. 54–55, 180.
44. This connection was undermined, at least, in the minds of all except the most diehard advocates of spontaneous generation, e.g., Haeckel. Lankester seemed reluctant to give up this connection, despite his staunch opposition to Bastian and to present-day spontaneous generation. This was evident in his 1876 translation of Haeckel's *History of Creation* (London: Macmillan).
45. Huxley, "On the Border Territory Between the Animal and the Vegetable Kingdoms" (1876), in *Discourses Biological and Geological, Essays* (New York: Appleton, 1925), pp. 165–200, quotation on p. 186.
46. [John M'Kendrick], "Obituary of John Hughes Bennett," *BMJ* 2: 473–478 (1875).
47. Virchow, "Standpoints in Scientific Medicine" (1877), in Leland J. Rather, trans. & ed., *Disease, Life and Man* (Stanford: Stanford University Press, 1958), pp. 146–147.
48. Schwann's first well-known experimental work in Müller's lab, three years before his *Mikroscopische Untersuchungen*, was an attempt to disprove the possibility of spontaneous generation. Recall that Bennett specifically opposed spontaneous generation even as late as early 1863 in his reply to Beale on that point.
49. Indeed, if his vitalistic interpretation were removed, Beale's own theory of "bioplasts" or "particles of germinal matter" reorganizing to form new cells or pathological structures in many ways resembles the heterogenesis that he so staunchly opposed.
50. Dr. William Cadge, Norwich to ?, 2 Oct. 1875 (Bennett Papers, Box Gen2007/4, unsorted, Edinburgh University Archives) letter about writing obituaries for two Edinburgh professors whose deaths he witnessed.
51. Anon., "Obit. of John Hughes Bennett," *Edinburgh Courant*, 27 Sept. 1875.
52. William J. O'Connor, *Founders of British Physiology* (Manchester: Manchester University Press, 1988), p. 97.
53. See Helena Curtis, *Invitation to Biology* (New York: Worth, 4th ed. 1985), p. 59, where Virchow's claim is embedded in the discussion of the origin of life; also W. G. Whaley, *Principles of Biology* (New York: Harper and Row, 3rd ed. 1964), p. 40.
54. See also Emil DuBois-Reymond, "Leibnizische Gedanken in der Neueren Naturwissenschaft," *Monatsberichte der Königlich Preussischen Akademie der Wissenschaft (Berlin)* July 1870: 835–854, a paper read on 1 July 1870

that bears striking resemblances in its discussion to Huxley's Liverpool address being written at that time.

55. Darwin to Hooker, 12 July 1870, *DCC* #7273, *MLCD*, v. 1: 321–322.
56. Stephen Brush, "Brownian Movement from Brown to Perrin," *Archives of History of the Exact Sciences* 5: 1–36 (1968); p. 2.
57. Anon., "Motion of Microscopic Granules," *MMJ* 5: 81–83.
58. C. C. Pode and E. Ray Lankester, "The Development of Bacteria in Organic Infusions," *PRSL* 21: 349–358 (1873).
59. Farley, for example, does not even mention Dallinger, or these experiments, in his book. Neither does Glenn Vandervliet (See *Microbiology and the Spontaneous Generation Debate During the 1870s* [Lawrence, Kans.: Coronado Press, 1971]). The only accounts I have seen in older literature on Dallinger's role are in Alison E. Adam, "Spontaneous Generation in the 1870s: Victorian Scientific Naturalism and its Relation to Medicine," Ph.D. diss., Sheffield Hallam University, 1988, and in John Crellin, "The Problem of Heat Resistance of Microorganisms in the British Spontaneous Generation Controversies of 1860–1880," *Medical History* 10: 50–59 (1966).
60. Most of this paragraph is based on J. W. Haas, "Late Victorian Gentlemen of Science: William Dallinger and the Wesley Scientific Society," paper presented 13 Sept. 1994 at the Royal Institution, London. See also Haas, "The Reverend Dr. William Henry Dallinger, F.R.S: Forgotten Victorian Microbiologist and Defender of the Faith," in *Proceedings of John Ray Conference, 18–21 March 1999* (in preparation).
61. See Dale Johnson, "Popular Apologetics in Late Victorian England: The Work of the Christian Evidence Society," *Journal of Religious History* 11: 558–577 (1981).
62. Haas, "Late Victorian Gentlemen of Science," p. 1.
63. See Samuel Thompson, *Essays Tending to Prove Animal Restoration* (Newcastle, 1830). My thanks to Jack Haas for this reference.
64. W. H. Dallinger and John Drysdale, "Researches in the Life History of a Cercomonad: A Lesson in Biogenesis," *MMJ* 10: 53–58 (1873); "Further Researches into the Life History of the Monads," *MMJ* 10: 245–249 (1873); 11: 7–10, 69–72, 97–103 (1874); 12: 261–269 (1874); 13: 185–197 (1875).
65. Dallinger and Drysdale, "Researches in the Life History of a Cercomonad," p. 53.
66. Ibid.
67. For more details on Dallinger, see his obituary in *PRSL B* 82: iv–vi (1910).
68. Dallinger and Drysdale, "Researches in the Life History of a Cercomonad," p. 57.
69. Ibid., p. 58.

70. Dallinger to Darwin, 10 Jan. 1876, *DCC* #10352, Darwin papers, APS Library, Philadelphia.
71. Darwin to Dallinger, n.d. (shortly after 10 Jan. 1876), *DCC* #10354, Tyndall papers, RI.CGlu/3.
72. Dallinger seems to have enclosed Darwin's letter when he first wrote to Tyndall, as that letter remains in the Tyndall archive.
73. Tyndall to Dallinger, 6 Apr. 1876, Tyndall papers 7/F 12.1.
74. Dallinger was surely sketching out the implications of Tyndall's suggestion, at least in theory, in his article "Recent Researches into the Origin and Development of Minute and Lowly Life-forms, with a Glance at the Bearing of These on the Origin of Bacteria," *Nature* 16: 24 (1877).
75. It is worth noting that Bastian did not find the heat resistance of the "sporules" convincing. He observed that these experiments only began to be widely cited after "Tyndall's contradictory experiments [fall 1875–spring 1876 vs. fall 1876] had in some way to be explained." He pointed out that everyone had known since at least 1862 that a temperature insufficient to kill spores of fungi in the dry state was more than sufficient to kill them in a moist state. Thus, reports that spores had survived 121° C of dry heat were not news. See Bastian, "On the Conditions Favouring Fermentation . . .," read at Linnean Society 21 June 1877, *Journal of the Linnean Society* 14: 1–94 (1879), pp. 76–79.
76. Pauline Mazumdar, *Species and Specificity: An Interpretation of the History of Immunology* (Cambridge: Cambridge University Press, 1995), pp. 39–40.
77. See Harris Coulter, *Divided Legacy. Part IV, Twentieth Century Medicine: The Bacteriological Era* (Berkeley: North Atlantic, 1994), pp. 181–211.
78. See T. H. Huxley, "On Penicillium, Torula, and Bacteria," *QJMS* 10: 355–362 (1870); E. Ray Lankester, "On a Peach-coloured Bacterium, *Bacterium rubescens*," *QJMS* 13: 408–425 (1873); and Joseph Lister, "A Contribution to the Germ Theory of Putrefaction and Other Fermentative Changes, and to the Natural History of Torulæ and Bacteria," read on 7 Apr. 1873, published in *Transactions of the Royal Society of Edinburgh* 27: 313–344 (1876), and Lister, "A Further Contribution to the Natural History of Bacteria and the Germ Theory of Fermentative Changes," *QJMS* 13: 380–408 (1873).
79. Huxley to Hooker, 20 July 1870, Huxley papers 2: 157–158.
80. Huxley to Hooker, 28 July 1870, Huxley to Hooker volume pp. 106–107, Hooker papers, Kew. The reason why Huxley (and others) misinterpreted what he saw has been discussed by Mary P. English in *Mordecai Cubitt Cooke: Victorian Naturalist, Mycologist, Teacher and Eccentric* (Bristol, U.K.: Biopress Ltd., 1987), pp. 141–142.
81. Hooker to Huxley, 30 July 1870, Huxley papers 3: 134–135.

82. Hooker to Huxley, 13 Aug. 1870, Huxley papers 3: 136–137. Hooker compared them in a somewhat nationalistic tone to the discoveries of the British mycologist Rev. W. F. Berkeley: "I have always fancied that it was rather brains and boldness, than eyes or microscopes that the mycologists wanted, and that there was more brains in Berkeley's crude discoveries than in the very best of the French and German microscopists' verifications of them, who filch away the credit of them from under Berkeley's nose, and pooh-pooh his reasoning."
83. This is also consistent with Huxley's position on "Xenogenesis" (his name for heterogenesis) in his 1870 Liverpool address. See T. H. Huxley, "Biogenesis and Abiogenesis," pp. 232–274 in *Discourses Biological and Geological,* pp. 261–262.
84. Bastian, *Beginnings of Life,* v. 1, p. xiv. John Farley (*Spontaneous Generation,* pp. 147–149) emphasized the importance of the solidification of a "law of genetic continuity" during the 1870s for the ultimate rejection of spontaneous generation. My analysis shows that Bastian felt he had an argument that made those new discoveries irrelevant to his case.
85. Bastian, *Beginnings of Life,* v. 1, p. xiv.
86. Bastian, *The Brain as an Organ of Mind* (London: Kegan Paul, 1880), p. 5.
87. Ibid.
88. Ibid., p. 6.
89. Roberts, "Studies on Abiogenesis," *PTRSL* 164: 457–477 (1874), p. 465.
90. See E. Ray Lankester, "The Pleomorphism of the Schizophyta," *QJMS* 26: 499–505 (1885). The image of *Bathybius* as a primordial protoplasmic plexus out of which more differentiated life forms in some way evolved was one of the characteristics that had such appeal that Haeckel and Lankester had difficulty letting go of "Huxley's brainchild" as Huxley himself had done in 1875. This idea was entangled with the pleomorphism, still popular with Lankester in 1879, but as objectionable to Huxley as his Urschleim creation, and this persistent entanglement may explain why Huxley was reported at the BAAS meeting that year to be "still sore under the mention of *Bathybius.*"
91. Louis Pasteur, *Etudes sur la Bière* (1876), *Oeuvres,* v. 5, p. 79; in English trans., *Studies on Fermentation* (London: Macmillan, 1879), p. 92.

6 Germ Theories and the British Medical Community

1. This chapter essentially seeks to take up some aspects of the story told by Margaret Pelling in *Cholera, Fever and English Medicine, 1825–1865* (Oxford: Oxford University Press, 1978) at the point where she left off, around 1865.

2. On the variety of "germ theories" see Nancy Tomes and John Harley Warner, "Introduction to Special Issue on Rethinking the Reception of the Germ Theory of Disease," *Journal of the History of Medicine and Allied Sciences* 52: 7–16 (1997) and Terrie Romano, "The Cattle Plague of 1865 and the Reception of 'the Germ Theory' in Mid-Victorian Britain," *Journal of the History of Medicine and Allied Sciences* 52: 51–80 (1997). See also William Bynum, "The Evolution of Germs and the Evolution of Disease: Some British Debates, 1870–1900," and Mick Worboys, "'Deeper than the Surface of the Wound': Lister, Surgeons and Germs in Britain, 1867–1885," both papers read at Dibner Institute conference, "Pasteur, Germs and the Bacteriological Laboratory," Cambridge, Mass., Nov. 1996. John Farley has shown that the conclusion that parasitic worms are spread by eggs and not spontaneously generated was not seen as any kind of general support for the germ theory of disease at this time. The two subjects, he demonstrates, were not seen as related to one another via any broad parasitic theory of disease until much later, well after 1880. See John Farley, "Parasites and the Germ Theory of Disease," in Charles Rosenberg and Janet Golden, eds., *Framing Disease* (New Brunswick, N.J.: Rutgers University Press, 1992), pp. 33–49. Jerry Gaw also discusses some of the variety of views in Britain, explaining specifically the reception of Lister's ideas on antiseptic surgery, in his "'A Time to Heal': The Diffusion of Listerism in Victorian Britain," *Transactions of the American Philosophical Society* 89 (part 1): 1–173 (1999). Gaw describes some of the editorial positions of the *Lancet* and *BMJ* that I consider here; however, his pro-Lister bias sometimes renders his analysis almost Whiggishly ahistorical in its inability to comprehend why opposition to the germ theory could be a reasonable, intelligent view at that time. See page 68 on James Wakley, for example.
3. See John Hughes Bennett, "Reports of Societies, B.A.A.S. Annual Meeting, Bath, September 1864: The Physiological Aspect of the Sewage Question," *BMJ* 2: 556–558 (12 Nov. 1864). For more on discussions surrounding the "sewage question" see Christopher Hamlin, *A Science of Impurity: Water Analysis in Nineteenth Century Britain* (Bristol: Adam Hilger, 1990), and Hamlin, "Providence and Putrefaction: Victorian Sanitarians and the Natural Theology of Health and Disease," in Patrick Brantlinger, ed., *Energy and Entropy* (Bloomington: Indiana University Press, 1989), pp. 93–123.
4. See Lloyd Stephenson, "Science Down the Drain," *Bulletin of the History of Medicine* 29: 1–26 (1955).
5. In the fall of 1877, a sharp dispute broke out between Burdon Sanderson and Tyndall, continuing into 1878, over exactly what the germ theory meant. The details are beyond the scope of this chapter, but reveal how varied and ambiguous the "germ theories" were, even well after the sup-

posed victory of germ theory over Bastian. One treatise appearing at this time explicitly referred to *The Germ Theories of Infectious Diseases* (by John Drysdale, London: Balliere, Tindall and Cox, 1878, my emphasis). Some of Burdon Sanderson's arguments featured interesting reincarnations of "histological molecules." See John Burdon Sanderson, "Bacteria," *Nature* 17: 84–87 (1877) and John Simon to Burdon Sanderson, 22 July 1877 (UCL archives) concerning an early draft of this paper; also Burdon Sanderson, *A Lecture on the Germ Theory and on the Doctrine of Contagium Vivum* (London: privately printed, 1878).

6. The historical parallels between this epidemic and the British beef scare of 1996 due to "mad cow disease" are striking in many ways, particularly the arguments against contagionism by those concerned about the economic devastation that might result from mass-slaughter or quarantine.
7. He was the father of E. Ray Lankester who is featured much more prominently in this story. The son took over editorship of the *QJMS* in 1868.
8. The entire text of this address was printed in *BMJ* 2: 437–443 (28 Oct. 1865).
9. Ibid., p. 439.
10. Ibid., p. 440.
11. Ibid.
12. Ibid., pp. 440–441.
13. Ibid., p. 441.
14. See John Eyler, "The Conversion of Angus Smith: The Changing Role of Chemistry and Biology in Sanitary Science, 1850–1880," *Bulletin of the History of Medicine* 54: 216–234 (1980), esp. p. 219.
15. Robert Angus Smith, *Disinfectants and Disinfection* (Edinburgh: Constable, 1869), pp. 29–31.
16. Robert Angus Smith, "On Organic Matter in the Air," *Proceedings of the Literary and Philosophical Society of Manchester* 9: 67–69 (1869–1870); also J. B. Dancer, "Microscopical Examination of the Solid Particles Collected by Dr. Angus Smith from the Air of Manchester," *Proceedings of the Literary and Philosophical Society of Manchester* 10: 270–274 (1871). These studies were carried out in 1867–1868. When John Tyndall gave his "Dust and Disease" lecture in Jan. 1870, Smith felt he and Dancer had not been given credit for coming to the same conclusions earlier. See his "On Organic Matter," pp. 65–66. See also John Crellin, "Airborne Particles and the Germ Theory: 1869–1880," *Annals of Science* 22: 49–60 (1966).
17. Joseph Lister, "Address in Surgery," *BMJ* 2: 225–233 (1871), p. 227.
18. Christopher Lawrence and Richard Dixey, "Practicing on Principle: Joseph Lister and the Germ Theories of Disease," in *Medical Theory, Surgical Practice,* ed. Christopher Lawrence (London: Routledge, 1992), pp. 153–215.

19. Ruth Barton, "'An Influential Set of Chaps': The X Club and Royal Society Politics 1864–85," *British Journal for the History of Science* 23: 53–81 (1990), p. 58. Of course, Busk and Huxley had had medical training early in their careers.
20. Frank M. Turner, "John Tyndall and Victorian Scientific Naturalism," in William H. Brock, N. D. McMillan, and R. C. Mollan, eds., *John Tyndall: Essays on a Natural Philosopher* (Dublin: Royal Dublin Society, 1981), pp. 176–177.
21. See Henry Bence Jones, *Lectures on Applications of Chemistry and Mechanics to Pathology and Therapeutics* (London: John Churchill, 1867), pp. 221–222, where Bence Jones discusses Pasteur's broad claims about external germs and concludes: "At present, notwithstanding M. Pasteur's high authority, the possibility of the production of infusoria without germs entering the body must be considered an open question" (p. 222). Bence Jones, while a "contingent contagionist" over his views on the Cattle Plague, actually *opposed* quarantining of diseased animals, joining "the landed interests, perhaps reflecting his family connections with them." See John R. Fisher, "British Physicians, Medical Science and the Cattle Plague, 1865–66," *Bulletin of the History of Medicine* 67: 651–669 (1993), esp. p. 661. Alternatively, Darwin's London physician, Dr. Henry Holland, a longtime believer in contagion via microorganisms, as well as in Ehrenberg's demonstration of air germs and Pasteur's victory over Pouchet, may have been a source that informed Tyndall's views. See Holland's 1839 essay "On the Hypothesis of Insect Life as a Cause of Disease," published in his *Medical Notes and Reflections* (London: 1842); also Holland to Darwin, 15 Jan. 1862, Apr. 1862, *DCC* #3390 and 3490, in Darwin papers, APS Library.
22. Bence Jones to Tyndall, 14 July 1870, Tyndall papers, tss. p. 696. For more on Gull's comments on Tyndall's "Dust and Disease" lecture, see text.
23. Alison Adam, "Spontaneous Generation in the 1870s: Victorian Scientific Naturalism and Its Relation to Medicine," Ph.D. diss., Sheffield Hallam University, 1988, pp. 166–167. For more on Frankland, see Colin Russell, *Edward Frankland: Chemistry, Controversy and Conspiracy in Victorian England* (Cambridge: Cambridge University Press, 1996).
24. Tyndall, "Dust and Disease," in *Essays on the Floating Matter of the Air* (New York: Johnson Reprint, 1966), p. 2. Tyndall's studies on the nature of dust in the air were informed by J. B. Dancer's recent work on this topic.
25. Tyndall to Huxley, Sept. 1869, Huxley papers v. 8, f. 72–74. Tyndall did not publicly describe this event, or its effect on his thought, until Oct. 1876, in "Fermentation and Its Bearings on Surgery and Medicine," in *Essays on the Floating Matter of the Air*, pp. 260–261.

26. Tyndall diary entry, Sunday 20 Nov. 1869, Tyndall papers 2/C8–12, p. 441. From Budd, Tyndall may have learned of Lister's recent attack (8 Nov. 1869) on Bennett's claim that the air germs of contagionists were imaginary. See Lawrence and Dixey, "Practicing on Principle," p. 210.
27. *Lancet* 1: 163–164 (21 Jan. 1870); *BMJ* 1: 118–119 (28 Jan. 1870). Reviews of Tyndall's Friday night Royal Institution Lecture, "Dust and Disease."
28. [H. Lawson], editorial, *Scientific Opinion* 3: 378 (27 Apr. 1870). The editor indicated basically the same opinion again on 1 June 1870 (p. 481).
29. Pouchet to Tyndall, 8 June 1870, Tyndall papers, 27/F6.
30. Arthur E. Durham, "President's Address, Delivered at the Annual Meeting, July 24th, 1868," *Journal of the Quekett Microscopical Club* 1: 95–110 (1868–69), quotation on p. 110. Durham, in his presidential address of a year later, seemed fully willing to grant serious consideration to Bastian's arguments from "The Origin of Life" in the *BMJ,* that the possibility of spontaneous generation was fully in line with tenets such as evolution and the "correlation of forces." Bastian had not yet publicly taken credit for this work, which he did only in his 1872 book, *The Beginnings of Life.*
31. Turner, "John Tyndall and Victorian Scientific Naturalism," p. 177; also Turner, "Rainfall, Plagues, and the Prince of Wales," in *Contesting Cultural Authority* (Cambridge: Cambridge University Press, 1993).
32. Richard French, *Anti-Vivisection and Medical Science in Victorian Society* (Princeton: Princeton University Press, 1975), pp. 55–60.
33. Beale to Burdon Sanderson, 14 Nov. 1870, Burdon Sanderson papers, Mss. 179/1, University College Library, London. Burdon Sanderson resisted this invitation to side against Tyndall, being still on good terms with Tyndall at this time. Tyndall fully reciprocated Beale's feelings, as when in a letter to Bence Jones he remarked:

> That letter in the Times looks very like little Lionel: I wonder that those religious men take such pleasure in fluttering round the margin of misstatements, to use a mild word. The letter notwithstanding this tampering with the truth contains one or two good and amusing points. The bias of the Times in this matter is very clear, what a pity it is that there are no means of bringing the blush to the cheek of a newspaper, after the error which it espoused and the truth which it opposed have been demonstrated! Perhaps however such opposition fulfills a useful though unknown purpose, and that the best philosophy of life is to accept it as one of the ingredients of our intellectual life. Is it not a fact that the mechanical irritation of substances in themselves without nutrition, may set the bowels going, and thus end a fit of constipation? A tightness of the brain may be similarly removed by the irritation of the Times.

(Tyndall to Bence Jones, 13 Aug. 1871, Tyndall papers, tss. p. 805.)

34. On Thudichum's stature in the British medical and physiological communities of this period, especially thanks to the patronage of John Simon, see David L. Drabkin, *Thudichum, Chemist of the Brain* (Philadelphia: University of Pennsylvania Press, 1958).
35. Thudichum, "The Relation of Microscopic Fungi to Great Pathological Processes, Particularly to the Process of Cholera," Royal Microscopical Society, June 10, 1868, *MMJ* 1: 14–27 (1869), pp. 26–27. Thudichum was responding to the cholera fungus theory of Hallier of Jena, published in 1867.
36. Hamlin, "Providence and Putrefaction," p. 99. Hamlin notes that Thudichum's progress was slow, however, so that by 1873 or so Simon's enthusiasm had shifted more in the direction of his other protégé's (Burdon Sanderson's) work, about the significance of bacteriology.
37. Joseph G. Richardson, *A Handbook of Medical Microscopy* (Philadelphia: J. B. Lippincott, 1871), p. 104.
38. Ibid., pp. 196–197.
39. Cited in "Dust and Disease," *BMJ* 2: 660 (17 Dec. 1870).
40. Anon., "A Word With Professor Tyndall," *BMJ* 2: 632 (10 Dec. 1870).
41. Ibid.
42. Ibid.
43. Richard Quain, "An Address Delivered at the Pathological Society of London," *BMJ* 1: 98–99 (1869).
44. Tyndall, *Dust and Disease* (London: 1872; New York: Johnson Reprint, 1966), with the letter from Lewes and Tyndall's apologia on pp. 38–43.
45. Ibid., p. 42.
46. Ibid., p. 18.
47. For explicit reactions of offense to Tyndall's haughtiness, see *Lancet* 16 Apr. 1870, pp. 555–556; 30 Apr. 1870, p. 626; and *BMJ* 10 Dec. 1870, p. 632; 17 Dec. 1870, p. 660.
48. See Murchison entry in *DNB*, also Murchison's collection of newspaper clippings on this episode, from Aug. to Oct. 1873 (Royal College of Physicians, Ms. 710).
49. "Typhoid Fever" letter to the editor, the *Times*, 9 Nov. 1874.
50. William Bynum, "Sir Joseph Fayrer: The Natural History and Epidemiology of Cholera," talk at University College, 3 Oct. 1994. See also "The Official Refutation of Dr. Robert Koch's Theory of Cholera and Commas" in the British *QJMS* 26: 303–316 (1885).
51. Carpenter to B. W. Richardson, 27 Aug. 1883, RCP, autograph letters collection. On Carpenter's, Richardson's, and Bennett's theoretical views on "zymosis" as the process of disease production, see Pelling, *Cholera, Fever and English Medicine*, pp. 142–143.
52. This analysis draws heavily on Pelling, *Cholera, Fever, and English Medi-*

cine. For her revision of the traditional Whiggish telling, based on Pasteur's account, that all medical men saw his and Liebig's fermentation theories as irreconcilable, see esp. pp. 141–145, 301–310. See also Fisher, "British Physicians," esp. p. 663, and chapter 5 of John M. Eyler's *Victorian Social Medicine* (Baltimore: Johns Hopkins University Press, 1979).

53. Sharpey to Bastian, 16 Nov. 1871, Wellcome Institute London, autograph letters collection.
54. See *Lancet* 2: 20–21, 60–61, 104–105, 175–176, 215–216 (1874).
55. Pettenkoffer's theory had important political implications about what kind of measures, enforced locally rather than by far-off federal bureaucrats, were most likely to control an epidemic. These implications never played out more clearly or dramatically than in the Hamburg cholera epidemic of 1892. See J. E. Strick, historical foreword to Thomas Brock, *Robert Koch: A Life in Medicine and Bacteriology* (Washington, D.C.: American Society for Microbiology Press, 1999). For a magisterial work on the interaction of science and bureaucracy in the 1892 Hamburg epidemic, see Richard Evans, *Death in Hamburg* (London: Oxford University Press, 1987).
56. William Bulloch, *History of Bacteriology* (Oxford: Oxford University Press, 1938), p. 184; Lewis to Bastian, 15 July 1873, Wellcome Institute London, autograph letters collection.
57. See D. Douglas Cunningham and Timothy Lewis, "Scientific Investigation into the Causes of Cholera," *Lancet* 1: 3–4, 38–41, 76–78 (2, 9, and 16 Jan. 1869); also anon. "Microscopic Organisms in Health and Disease: Review of Recent Works by Lewis and Cunningham," *BFMCR* 56: 348–362 (Oct. 1875). This reviewer pointed out that "Drs. Lewis and Cunningham are peculiarly fitted to pass a judgment on these points, since, besides their medical and biological training, they have worked under the great fungologist, DeBary, with the special object of becoming acquainted with microscopic vegetation. We may, therefore, conclude that they know, as far as may be known, what is and what is not a vegetable organism, which could hardly be asserted of all investigators who have attacked this difficult theme." p. 348. [Was this a barb at Tyndall?]
58. Norman Howard-Jones, *The Scientific Background of the International Sanitary Conferences, 1851–1938* (Geneva: World Health Organization, 1975), p. 37.
59. Carpenter, 13 Nov. and Bree, 14 Nov. 1874 letters in the *Times*. In one of the most blatantly, albeit nonchalantly, Whiggish accounts of this, Tyndall's first biographers describe these views as a "curious exhibition of unscientific thought." See *LWJT*, p. 197. This is only somewhat more extreme than all of the histories of this subject prior to Farley's 1977 book.
60. [Thomas Lauder Brunton], "Another Aspect of the Tyndall Typhoid Controversy," *Practitioner* 14: 62–67 (1875), quotation on p. 63.

61. Ibid., p. 65.
62. Ibid., pp. 65–66.
63. William Osler's biographer, for instance, says of the International Medical Congress in London in August 1881 that "much merriment was provoked in the lay press at the expense of microbes in general" (pp. 190–191). A piece at that time in the *BMJ* on "Bacteria" sarcastically reported: "A Boston professor said, after a late lecture, to one of his class: 'The views in regard to bacteria are now so conflicting that I do not understand them.' 'Why Professor,' said the student, '*that is what we all say.*'" See *BMJ* 2: 320 (1881), emphasis in original. Perhaps in the humor one may see an embarrassed confession of confusion in having to back away from Bastian after the recent successes of Koch and Pasteur.
64. Anon., "The Origin of Life," *Lancet* 2: 50–51 (9 July 1870).
65. Ibid., p. 51.
66. Ibid.
67. Anon., "Review of *The Beginnings of Life*," *Practitioner* 9: 291–294 (1872), p. 291.
68. Ibid.
69. Ibid., p. 293.
70. Ibid., p. 292.
71. *Lancet* 2: 528–529, 563–564 (1872).
72. Anon., "The Evolution of Life: Review of Dr. Bastian's Book entitled 'The Beginnings of Life,' Reprinted from the Med. Press and Circular, September 25, 1872" (London: Ballière, Tindall and Cox, 1872), quotation from p. 15. This pamphlet version found in Linnean Society Library, Burlington House, London. Thanks to archivist Gina Douglas for help in locating this item.
73. Ibid., pp. 16–17.
74. Anon, "Bastian on Lowest Organisms," *BFMCR* 49: 448–451 (1872), quotation on p. 448.
75. Ibid., pp. 450–451.
76. Ibid., p. 451.
77. Charles Murchison, *On the Continued Fevers of Great Britain*, 2nd ed. (London: Macmillan, 1873), pp. xvi, 9–10.
78. Anon. editorial, London Daily Telegraph, 14 Aug. 1873.
79. Ibid.
80. It was at least five years after the young Darwinians had all gotten the message about spontaneous generation, for example, that rising young medical star William Osler began to doubt the doctrines of his teacher Bastian, in response to the new discoveries of Koch and other Continental bacteriologists. This was true in spite of Osler's acquaintance with Darwin and Dohrn, and full knowledge of the controversy from working in Burdon

Sanderson's lab at University College London. See Harvey Cushing, *Life of Osler* (London: Oxford University Press, 1940), v. 1, pp. 118–119, 190–191, 199–200, 215–216, 282.

81. The debate was reported verbatim in the *Lancet* 1: 166–167, 269–271, 337–342, 410–417 (1874).
82. Bastian's entire introductory paper was printed in the 10 April issue of *Lancet* 1: 501–509; also in *MMJ* 14: 65–79, 129–140 (1875).
83. The entire debate was reported as "Discussion on the Germ Theory of Disease" in *Transactions of the Pathological Society of London,* v. 26. Bastian's paper occupies pp. 255–284, Burdon Sanderson's remarks, pp. 284–289.
84. The discussion of the second meeting occupies ibid., pp. 290–314.
85. This discussion occupies ibid., pp. 314–334.
86. Ibid., pp. 334–345. Bastian's response and summary was also published in the 15 May issue of *Lancet* 1: 682–685. Citations from the entire discussion will be from the *Transactions of the Pathological Society of London* version.
87. Ibid., p. 257.
88. Ibid., p. 259.
89. Beale, a prominent and outspoken vitalist, was a fierce opponent of spontaneous generation. Nonetheless, his theory bears a strong resemblance to heterogenesis, though he insisted his bioplasm particles were not the same as bacteria. Ironically, because he was so irascible, he was at times more ill-disposed toward Tyndall and Huxley than toward Bastian, despite Bastian's much more direct opposition to vitalism, followed out to the logical conclusion of spontaneous generation.
90. On their emerging differences, see Beale to Burdon Sanderson, 1 Nov. 1869 and Burdon Sanderson reply (n.d., but labeled "reply"), Burdon Sanderson papers, University College London, Ms. Add. 179/1. This was expressed publicly by Beale in his two-part critique of 1870: *Disease Germs: Their Supposed Nature* and *Disease Germs: their Real Nature* (London: Churchill). Second edition of the pair republished together as *Disease Germs* (London: Churchill, 1872).
91. Bastian, "Discussion on the Germ Theory of Disease," p. 265.
92. Ibid., p. 266.
93. Ibid., p. 267.
94. See French, *Anti-Vivisection.*
95. Marshall was professor of surgery, Schäfer started out as Burdon Sanderson's assistant and later became professor of physiology when Burdon Sanderson moved to Oxford. Ringer was professor of medicine and Thane Professor of Anatomy.
96. Physiological Society Minute Book, Ms. E-19, Box 9, Sharpey-Schafer papers, Contemporary Medical Archive Center, Wellcome Institute London;

E. Sharpey-Schafer, *History of the Physiological Society During its First Fifty Years* (Cambridge: Cambridge University Press, 1927). Because of not even being mentioned in Sharpey-Schafer's history, Bastian also disappears from subsequent histories patterned after it, such as W. J. O'Connor's *Founders of British Physiology.* This despite O'Connor's inclusion up until 1876 of anybody who had almost any interest in physiology, such as George Eliot. Indeed, the entire spontaneous generation controversy disappears in this book, even from the careers of some of those most involved, such as Burdon Sanderson and Huxley.

97. See Foster to Dyer, 23 July 1876 ("Sanderson has some new maggots in his head . . ."), Thistleton-Dyer Correspondence, v. 2: 18–18a, Royal Botanical Gardens, Kew; also Sharpey-Schafer's more muted version in *History of the Physiological Society,* pp. 23–24. Tyndall described Burdon Sanderson as having a "vague and wandering mind," but only said this in private. See chapter 7.
98. Bastian to University College London, 24 July 1876, College Correspondence File, University College Archives.
99. See Bastian reviewer's report of 28 July 1879 on paper by Bevan Lewis, RR.8.169, RS Archives.
100. See Youmans to Lockyer, 12 Apr. 1872, Lockyer papers, Exeter University Library; also the forthcoming book on the International Scientific Series by Michael Collie and Leslie Howsam.
101. *LLHS,* see esp. v. 2: 168–169 (Spencer to Bastian 25 Nov. 1878 and Bastian reply of 27 Nov.); also personal communication from Michael Collie 15 Oct. 1994 and 5 Feb. 1996. Bastian's book was published in the series in 1880.
102. Bastian to Owen, 28 Oct. 1876, Owen papers, BMNH 2: 324–326. It is interesting that Bastian had never before contacted Owen, despite Owen's early support for heterogenesis. Bastian saw himself as a Darwinian, and Owen had thus always been viewed as an unsavory companion. No regular exchange of letters developed, so it may be that Bastian still wanted to avoid the association of being considered in any way a client of Owen's patronage.
103. Physiological Society Minute Book, Ms. E-19, p. 21.
104. Ibid.
105. Burdon Sanderson to Schäfer, 27 Apr. 1877, 28 July 1877 and 16 Apr. 1878, Sharpey-Schafer papers B7/10, 12 and 14. For example, in the 28 July letter: "I write these words that we may all know that Lankester is a greater fool than we took him for and act accordingly." See also Wm. Sharpey to Schäfer, 27 July 1877 on Lankester's success in taking over the elementary biology course at University College London from Burdon San-

derson (Sharpey Schafer papers B.3/16, Wellcome Institute London. And on Lankester's view of this, see Lankester to University College London, 29 July 1877 and 30 Apr. 1878, Lankester papers, College Correspondence file 1876–1877 and 1878.

106. D. E. Allen, "The Biological Societies of London, 1870–1914: Their Interrelationships and their Responses to Change," *Linnean* 14: 23–38 (1988), p. 23.
107. Huxley and Martin's *Elementary Biology* text made a claim that stretched the facts extraordinarily, suggesting a felt need to still respond to the campaign in favor of spontaneous generation as late as the 1888 second edition. There they stated, "*All forms* of bacteria . . . are found to produce resting spores." T. H. Huxley and H. N. Martin, *A Course of Elementary Instruction in Practical Biology* (London: Macmillan, 1888), p. 409, italics mine. This statement was not in the first edition. This was neither accepted by bacteriologists then or now. Only a few species, predominantly those of today's genera *Bacillus* and *Clostridium,* are capable of this, and even in the earliest days of bacteriology, such a sympathetic Huxley supporter as E. Ray Lankester claimed that only *B. subtilis* and *B. anthracis* could produce spores. See H. J. Harris, lecture notes for Introductory Zoology course taught by E. Ray Lankester at University College London from 1886–1890, probably 1889, pp. 67–78. Ms. Add. 95, Lankester papers, University College Archives. The texts Lankester required for this course were Huxley's *Lessons on Physiology,* Huxley and Martin's *Elementary Biology,* Klein's *Histology,* and Quain's *Anatomy.*
108. See the class notes of his student William Halliburton (Wellcome Institute London, Ms. 2705). Halliburton had great respect for Bastian and his ideas, even when he concluded later in his career that Bastian was probably wrong about the germ theory and spontaneous generation (see *Lancet,* 7 June 1919). But Halliburton is the exception that proves the rule. He became a practicing physician, while Lankester, Foster, and other Huxley protégés trained the entire next generation of British biologists. Thus, by the time Bastian began to publish on spontaneous generation again after 1900, both Lankester and Schäfer could express their opposition from the prestigious "elder statesman" position of the BAAS presidential address (Lankester in 1906 and Schäfer in 1912).
109. Herzen to Bastian, 15 Dec. 1877; also see Giuseppe Colomba to Bastian, 27 July 1881 and Viguiér to Bastian, 24 Nov. 1881; all in Wellcome Institute London, autograph letters collection. Other Continental supporters in the 1880s included J. P. Béchamp and Marburg botanist Albert Wigand. See criticism of their views on abiogenesis by Koch supporter Carl Flügge in his *Ferments and Microparasites,* 2nd ed., English trans. by W. W. Cheyne (London: Sydenham Society, 1890), p. 82.

7 Purity and Contamination

1. James Friday, Roy MacLeod, and Philippa Shepherd, *John Tyndall, Natural Philosopher, 1820–1893* (London: Mansell, 1974), p. 10.
2. J. R. Friday, "A Microscopic Incident in a Monumental Struggle: Huxley and Antibiosis in 1875," *British Journal for the History of Science* 7: 61–71 (1974); John Farley, *The Spontaneous Generation Controversy from Descartes to Oparin* (Baltimore: Johns Hopkins University Press, 1977), esp. pp. 132–141; Alison E. Adam, "Spontaneous Generation in the 1870s: Victorian Scientific Naturalism and Its Relation to Medicine," Ph.D. diss., Sheffield Hallam University, 1988.
3. Evolutionists who got the message about Bastian and thus abandoned their earlier interest in spontaneous generation as at least a possibility include John Fiske, Frederick Barnard, Thomas Clifford Allbutt, William Halliburton, Alfred R. Wallace, and Henry Lawson. Only Gilbert Child and T. R. R. Stebbing seem never to have felt the need to recant. That the rest dropped their past associations with Bastian is emphasized by the fact that many of their biographers either made entirely no mention of Bastian and this rather dramatic controversy (Lockyer, Fiske, Hooker, Youmans), or mentioned only their later, negative opinions of Bastian and his work (especially Tyndall). Friday, MacLeod, and Shepherd (*John Tyndall, Natural Philosopher,* p. 10) have also taken Tyndall's account uncritically, and represented the whole revival of interest in spontaneous generation in the mid-nineteenth century as a direct result of "the development of high-powered compound microscopes." In this, as in so many other cases involving improvements in technology, historians and philosophers of science have shown that philosophical, religious, and political commitments were much more significant as sources of this interest, while the new microscopes at first actually caused *more* debate and confusion about what was actually being *seen* through them. See L. S. Jacyna, "John Goodsir and the Making of Cellular Reality," *Journal of the History of Biology* 16: 75–99 (1983) and Jacyna, "The Romantic Programme and the Reception of Cell Theory in Britain," *Journal of the History of Biology* 17: 13–48 (1984).
4. Farley, *Spontaneous Generation Controversy,* p. 86.
5. R. Owen, *On the Anatomy of Vertebrates,* v. 3, pp. 817–825; see also D. M. Knight, *Atoms and Elements* (London: Hutchinson, 1967), pp. 51–52, 104–126 on the debates in the Chemical Society, the invoking of Faraday by anti-Daltonians, and Tyndall's reaction.
6. Henry Bence Jones, *Life and Letters of Faraday* (London: Longmans Green, 1870), v. 2, pp. 276–279.
7. See *LLJDH,* v. 2, pp. 112, 359. (Hooker to Darwin, *DCC* #5971): "Tyndall believes he feels atoms, as firmly as St. Paul believed he saw Christ. I do

not say that atoms do not exist, but I rather suppose that they may be like minutes of time or inches of space or any other *purely arbitrary quantities.*" (Hooker to his son): At an X Club meeting once, "the discussion about Atoms waxed hot between Frankland and Tyndall, Huxley quietly giving occasional dagger thrusts to both, and especially to Tyndall, who held that he saw atoms visually in his mind's eye, and who on his inability to describe to Huxley what he saw, was met by H. with the rejoinder 'Ah, now I see myself; in the beginning was the Atom, and the atom *is* without form and void, and darkness *sits* on the face of the Atom!'"

8. Tyndall, *Faraday as a Discoverer* (London, 1868), reprint (New York: Thos. Crowell, 1961), p. 145.
9. See Henry Charlton Bastian, "Facts and Reasonings Concerning the Heterogeneous Evolution of Living Things," *Nature* 2: 170–177 (1870), esp. p. 175n. It is obvious here that Bastian also models his understanding of molecular forces in protoplasm very closely after Huxley's 1868 lecture, "The Physical Basis of Life." Re: Bastian's first contact with Owen, see Bastian to Owen, 28 Oct. 1876, Owen papers, BMNH 2: 324–325.
10. Henry Bence Jones, *Croonian Lectures on Matter and Force* (London: John Churchill, 1868), cited in Bastian's *Beginnings of Life,* v. 1, p. 62.
11. See P. L. Sawyer, "Ruskin and Tyndall: The Poetry of Matter and the Poetry of Spirit," in J. Paradis and T. Postlewait, eds., *Victorian Science and Victorian Values: Literary Perspectives* (New Brunswick, N.J.: Rutgers University Press, 1985), p. 235. Tyndall leaned heavily on the analysis of Friedrich Lange's *Geschichte des Materialismus* for the history of atoms in his 1874 Belfast Address. Interestingly, however, he chose to omit Lange's criticism of Pasteur and sympathy for Pouchet, and for Haeckel's view of *Bathybius* as evidence for spontaneous generation.
12. Bastian, "Facts and Reasonings," p. 175. It should be noted that Bastian's analogy with crystallization was not itself the subject of criticism by Huxley, Spencer, Darwin, and others. It was only when he claimed this possible for a wide *variety* of organisms much more complex than *Bathybius* (the first step to protoplasm, as still viewed by Huxley until 1875), or even bacteria, that the initial interest of these men turned to skepticism. See Darwin to Hooker, 12 July 1870, *DCC* #7273, Darwin papers, APS: "I cannot persuade myself that such a multiplicity of organisms can have been produced, like crystals, in Bastian's solutions." Nonetheless, Darwin still felt in Sept. 1873 (letter to Haeckel) that "if it [spontaneous generation] could be proved true, this would be most important to us." See also M. A. DiGregorio, *Charles Darwin's Marginalia* (New York: Garland, 1990), pp. 34–35. Edward Frankland of the X Club was still working with Bastian on experiments in Nov. 1870, and the two were given a Royal Society research grant for the work by then Secretary William Sharpey. Sharpey

was still on cordial terms with Bastian and showed respect for his theories in a letter of 16 Nov. 1871 (Wellcome Institute London, autograph letters collection; Royal Society, MC 9.129, 133).

13. On Owen's declining influence, see E. Richards, "A Political Anatomy of Monsters, Hopeful and Otherwise," *Isis* 85: 377–411 (1994), p. 392; also Alvar Ellegaard, *Darwin and the General Reader* (Chicago: University of Chicago Press, 1990), p. 51. Gordon Holmes has described Bastian as "a friend and later the literary trustee of . . . Herbert Spencer, and was one of the group that included Darwin, Russel Wallace, Huxley, and others which finally established the theory of evolution." (*The National Hospital, Queen Square, 1860–1948* [London: Livingstone, 1954], p. 39.)
14. Tyndall to Darwin (n.d., but c. 1858), Tyndall papers, tss. 2889. See also Tyndall, "On the Importance of the Study of Physics," lecture at the Royal Institution, reprinted in E. L. Youmans, ed., *Modern Culture: Its True Aims and Requirements* (London: Macmillan, 1867).
15. Roy MacLeod, "Science in Grub Street," *Nature* 224: 423–434 (1969), quotation on p. 432.
16. Huxley, Frankland, Busk, and Sharpey all initially worked with Bastian as "witnesses" (in Shapin and Schaffer's sense), showing eager interest in what he might actually produce, and Huxley himself during this period was experimenting on the question and continued to do so even after he lost confidence in Bastian, during the summer of 1870.
17. Tyndall, "Dust and Disease," in *Essays on the Floating Matter of the Air* (New York: Johnson Reprint, 1966), p. 33.
18. Tyndall to Bastian, 25 Jan. (n.y., but must be 1870), Tyndall papers, T6/C10, tss. p. 149. Note that all of Bastian's books during the 1870s controversy were published by Macmillan, the London publisher who was fully committed to the "young guard" of science, e.g., by beginning the journal *Nature* in 1869. See Roy MacLeod, "Macmillan and the Young Guard," *Nature* 224: 435–461. E. L. Youmans, the American equivalent of Macmillan, was also good friends with Bastian as well as Tyndall and Huxley, and very supportive with publicity (Youmans to Bastian, 16 Apr. 1874, Wellcome Institute London autograph letters collection), courting Bastian to write a book for the International Scientific Series in 1875.
19. Tyndall to Huxley, 14 Sept. and 3 Oct. 1870, Huxley papers, 8: 82, 84. After Burdon Sanderson seemed unwilling to condemn Bastian's work publicly and made statements viewed by many as supporting Bastian, Tyndall changed his assessment. Though to Burdon Sanderson's face he said:

> My own position in these discussions has, I have reason to know, been far more influenced by your being cited against me than by the more direct antagonism of Dr. Bastian. Nothing gives me so much discomfort as un-

> certainty, and it is with the view of getting rid of this that in the paper presented to the Royal Society last Thursday I have sought clearly to define our relative positions. I have in the first place been obliged to refer to the arguments which Dr. Bastian has founded upon your experiments, and which he reproduced at the Royal Society a short time ago; and I dare not pass over without remark your recent utterance before the Society of Medical Officers. The occasion appeared to me to be a very grave one, and I have acted on this supposition.

(UCL Library, Burdon Sanderson papers, 179/1). By contrast, to his friend Emil DuBois-Reymond, he said: "Sanderson is clever and industrious, but he possesses a vague and wandering mind. He gave a lecture to the Officers of Health here some time ago in which I felt called upon to make some remarks which I fear were not agreeable to him. I send you a brief abstract of the remarks: they will probably appear at greater length in the Philosophical Transactions" (Tyndall papers 24/B8.16).

20. Tyndall to Barnard, 10 Dec. 1873, X mss. collection, X973/c72/f, Columbia University Library. I am indebted to Michael Collie for this reference. Tyndall was replying to Barnard's article "The Germ Theory of Disease and its Relations to Hygiene," which I have discussed in chapter 5.
21. On Tyndall's hero-worship, and the inevitably connected tendency to make villains out of those who opposed him or his heroes, see Sawyer, "Ruskin and Tyndall," esp. pp. 225–227.
22. Tyndall to Huxley, 15 Oct. 1876, Tyndall papers, 14/C9.80.
23. Tyndall to Mme. Novikoff, 15 Oct. 1876, Tyndall papers, tss. p. 941.
24. Pasteur to Tyndall, 9 May 1870, Tyndall papers 2/C11: 478.
25. Tyndall to Bence Jones, 17 May 1870, Tyndall papers tss. p. 695; Review, "Pasteur's Researches on the Diseases of Silkworms" in *Nature* 2: 181–183 (7 July 1870, immediately preceding part 2 of Bastian's long three-part article, first announcing and describing Bastian's experiments on spontaneous generation).
26. Pasteur to Tyndall, 12 Nov. 1874, Tyndall papers 27/C1.3.
27. Tyndall to Pasteur, 7 Feb. 1876, Tyndall papers, photocopy of original from Bibliothèque Nationale (BN hereafter).
28. Pasteur to Tyndall, 8 Feb. and 9 Feb. 1876, Tyndall papers 27/C1.5 and 1.4.
29. Tyndall to Darwin, 2 Feb. 1876, *DCC* #10377, Darwin papers, APS; Tyndall papers tss. p. 2839.
30. Darwin to Tyndall, 4 Feb. 1876, *DCC* #10379, Darwin papers, APS; Tyndall papers tss. p. 2850.
31. Tyndall, "Optical Deportment of the Atmosphere in Relation to Putrefaction and Infection," *PTRSL* 166: 27–74 (1876), also in Tyndall, *Essays on the Floating Matter of the Air*, pp. 108–111.

32. Dallinger to Darwin, 10 Jan. 1876, *DCC* #10352, Tyndall papers RI.CGlu/3.
33. See Tyndall, "Optical Deportment of the Atmosphere," p. 126. See also Dallinger's articles at this time, "Professor Tyndall's Experiments on Spontaneous Generation and Dr. Bastian's Position," *Popular Science Review* 15: 113–127 (1876) and "Practical Notes on Heterogenesis," *Popular Science Review* 15: 338–350. Note that these articles were published in one of Henry Lawson's journals.
34. William Roberts, "A Word on the Origin of Bacteria, and on Abiogenesis," *BMJ* 1: 282–283 (1876), p. 283.
35. Ibid.
36. And Adam has shown ("Spontaneous Generation in the 1870s," p. 114) that in the British debate, no one experimental "blow" was ever decisive in resolving the issues at stake.
37. Tyndall to Pasteur, 1 July 1876, Tyndall papers, loose photocopy of BN original.
38. Tyndall to Pasteur, 21 Aug. 1876, Tyndall papers, loose photocopy of BN original. The degree of success Tyndall had in eventually convincing Pasteur that Bastian was a true enemy can be seen in similar remarks that Pasteur afterward made about Bastian. Osler witnessed an exchange between the two men in 1881: "The great Pasteur produced a sensation by confessing that his ignorance of English and German had prevented his following the arguments of the previous speakers; and then by exclaiming in reply to Dr. Bastian, who, he was told, held that micro-organisms may be formed by heterogenesis of the tissues: 'Mais, mon Dieu, ce n'est pas possible,' and without advancing any argument then sat down. The eminent man for the moment seemed unable to realize the possibility of intelligent dissent from his assertion" (Harvey Cushing, *Life of Osler* (London: Oxford University Press, 1940), v. 1, pp. 190–191). Pasteur was very consistent in this approach to anyone he viewed as an antagonist. See G. L. Geison, *The Private Science of Louis Pasteur* (Princeton: Princeton University Press, 1995).
39. Tyndall to Pasteur, 5 Sept. 1876, Tyndall papers, photocopy of BN original.
40. See Stella Butler, R. H. Nuttall, and Olivia Brown, *The Social History of the Microscope* (Cambridge: Whipple Museum of History of Science, 1986).
41. Tyndall, "Optical Deportment of the Atmosphere," p. 78.
42. Tyndall to Pasteur, 16 Feb. 1876, Tyndall papers, loose photocopy of BN original.
43. Adam, "Spontaneous Generation in the 1870s," p. 98.
44. See criticisms of this sort even from a doctor in many ways sympathetic to Tyndall: T. Maclagan, letter to editor, *Lancet* 1: 295–296 (19 Feb. 1876).
45. Lionel Beale, "Spontaneous Generation and the 'Searching Beam,'" *BMJ* 1: 254 (1876).
46. Roberts, "Studies on Abiogenesis," *PTRSL* 164: 457–477, p. 472.

47. Tyndall to Huxley, 20 Feb. 1876, Huxley papers 8: 189.
48. Huxley to Tyndall, 7 Oct. 1876, Huxley papers 8: 192. In a letter to E. L. Youmans a few months before this, Tyndall had stated, "I wish Bastian had permitted me to treat him tenderly. I expressed this wish to himself, but, as Huxley says, tenderness to him is sure to be misinterpreted." See John Fiske, *Edward Livingston Youmans, Interpreter of Science* (New York: Appleton, 1885), p. 320.
49. Ruth Barton, "'An Influential Set of Chaps': The X Club and Royal Society Politics 1864–85," *British Journal for the History of Science* 23: 53–81 (1990), pp. 53–54, 59–74.
50. Bastian to Lockyer, 18 Jan. 1876, Lockyer papers, Exeter University Library.
51. Royal Society Register of Papers (Ms. 421), RS Archives. Only Huxley's review of 10 Apr. 1876 was preserved in the RS Archives (Reviewers' Reports v. 7, f. 496). Tyndall's paper appeared as "Optical Deportment of the Atmosphere."
52. Royal Society Register of Papers. According to Gerald Geison, Rolleston, who had been teaching a course at Oxford since 1860, had "almost certainly developed his course under the inspiration and guidance of Huxley, who had helped him to secure his chair in the first place." See *Michael Foster and the Cambridge School of Physiology* (Princeton: Princeton University Press, 1978), p. 135.
53. The fact that no written trace remains of any of the reports on Bastian is also somewhat suggestive. The only indication is an annotation next to Sorby's name as reviewer, in the margin of the record book, "report to Mr. Stokes by Sorby." On Henry C. Sorby, see Samuel Alberti, "Field, Lab, and Museum: The Practice and Place of Life Science in Yorkshire, 1870–1900" (Ph.D. diss., Universities of Leeds and Sheffield, forthcoming 2000); also Norman Higham, *A Very Scientific Gentleman: The Major Achievements of Henry Clifton Sorby* (London: Pergamon, 1963); and Michael J. Bishop, "New Biographical Data on Henry Clifton Sorby (1826–1908)," *Earth Sciences History* 3: 69–81 (1984). George G. Stokes was a devoutly religious man, who had outspokenly opposed evolution. He suggested in his 1869 BAAS presidential address that Huxley's "Physical Basis of Life" address was gravely mistaken in suggesting that the boundary between living and nonliving could ever be understood by science.
54. Huxley to Tyndall, 18 Nov. 1876, Huxley papers 8: 197.
55. Huxley to Tyndall, 14 Jan. 1877, Huxley papers 8: 200.
56. Tyndall to Pasteur, 17 Jan. 1877, Tyndall papers, photocopy of BN original.
57. Bastian to Huxley, 25 Jan. 1877, Huxley papers 10: 242.
58. Huxley to Bastian, 26 Jan. 1877, Huxley papers 10: 243.

59. Bastian, "On the Conditions Favouring Fermentation and the Appearance of Bacilli, Micrococci and Torulae in Previously Boiled Fluids," presented 21 June 1877, published in *Journal of the Linnean Society of London* 14: 1–94 (1879). The Linnean Society was where Bastian's very first paper had first been presented and published.
60. Bastian, "Researches Illustrative of the Physico-chemical Theory of Fermentation, and of the Conditions Favouring Archebiosis in Previously Boiled Fluids," RS Archives AP.58.8. Published in abstract in *Nature* 14: 220–222 (1876).
61. Tyndall, "Further Researches on the Deportment and Vitality of Putrefactive Organisms from a Physical Point of View," *PTRSL* 167: 149–206 (1877); also in Tyndall, *Essays on the Floating Matter of the Air.*
62. Darwin to Romanes, 23 May 1877, *DCC* #10971, APS Library, Carroll #513.
63. This allowed him to exploit certain weaknesses that had recently come to light with "filth," as in the Aug.-Sept. 1873 typhoid epidemic in Marylebone. The *Daily Telegraph* (14 Aug.) puzzled over how the "filth" needed as a source of the typhoid poison could have gotten into such an upper middle-class neighborhood where many upstanding professionals, including many of London's best known doctors, had their homes.
64. Tyndall to Darwin, 23 Feb. (1871?), *DCC* #7508, Tyndall papers tss. p. 2809.
65. *LWJT*, p. 320 (italics mine).
66. Cohn was an old friend of Tyndall (Cohn to Tyndall, 22 Apr. 1872, Tyndall papers 29/E3.B1) and had a theoretical interest as well in disproving heterogenesis, since it undermined his major project of producing a taxonomy of stable bacterial species even more than pleomorphic theories did. See William Bulloch, *History of Bacteriology* (London: Oxford University Press, 1938), pp. 184, 188, 192–210.
67. For a well-developed exposition of the experiments on the heat-resistant spore question, see G. Vandervliet, *Microbiology and the Spontaneous Generation Debate During the 1870s* (Lawrence, Kans.: Coronado Press, 1971). This study also illustrates the total emphasis on the experiments as the whole story, however, and thus totally overlooks the issue of the experimenter's regress so clearly implicit *before* the closure via acceptance of heat-resistant spores had occurred.
68. Note how the most common expressions we have inherited from Huxley, Tyndall, and Pasteur (the winners who wrote the histories) guide us to see what's happening in their terms.
69. Tyndall, "On the Optical Department of the Atmosphere in Reference to the Phenomena of Putrefaction and Infection," *PRSL* 24: 176 (1876).

70. Ibid., p. 174. The full-length version of Tyndall's "500 chances" article appeared in the *PTRSL* 166: 27–74 (1876).
71. The morally loaded nature of the supposedly objective terms of the debate, such terms as "clean," "contaminated," and "impurity," was first noted by Lloyd Stevenson in his classic article "Science Down the Drain: On the Hostility of Certain Sanitarians to Animal Experimentation, Bacteriology, and Immunology," *Bulletin of the History of Medicine* 29: 1–26 (1955), pp. 8–9.
72. Farley, *Spontaneous Generation Controversy*, p. 132.
73. Ibid., pp. 134–135, 140–141.
74. Tyndall, "Preliminary Note on the Development of Organisms in Organic Infusions," *PRSL* 25: 503–506 (1877), p. 504.
75. On the condition of the Royal Institution labs, see Donovan Chilton and Noel Coley, "The Laboratories of the Royal Institution in the Nineteenth Century," *Ambix* 27: 173–203 (1980).
76. Tyndall's laboratory notes on this series begin 7 Oct. 1876, Tyndall papers, notebook 2/E5, pp. 81ff. The first public notice Tyndall gave of his awareness of spores was his mention of the companion article in Cohn's journal by Koch, in a lecture at Glasgow on 19 Oct. 1876 (Tyndall, *Essays on the Floating Matter of the Air,* pp. 267–270). He also mentioned Koch's paper in a letter to Pasteur of 27 Oct. (Tyndall papers 27/C1.11).
77. Bastian, "On the Conditions Favouring Fermentation," pp. 70–71.
78. Pasteur to Tyndall, 26 Aug. 1876, Tyndall papers 27/C1.7.
79. Tyndall to Pasteur, 27 Oct. 1876, Tyndall papers, photocopy of BN original.
80. Bastian, "The Fermentation of Urine and the Germ Theory," *Nature* 14: 309–311 (1876); also in *Lancet* 2: 248–249 (1876).
81. Tyndall to Pasteur, 13 Nov. 1876, Tyndall papers, photocopy of BN original.
82. Tyndall to Pasteur, 15 Nov. 1876, Tyndall papers, photocopy of BN original.
83. Pasteur to Tyndall, 1 Dec. 1876, Tyndall papers 27/C2.14. The letter to Bastian was not found with this ms. I have, however, found a manuscript of the correct date (23 Sept. 1876) and appropriate contents from Pasteur to an anonymous recipient, which seems to be this letter to Bastian. It is in the Pasteur papers (ms. 5127) at the Wellcome Institute London.
84. Pasteur to Tyndall, 13 Jan. 1877, Tyndall papers 27/C3.20.
85. Tyndall to Pasteur, 17 Jan. 1877, Tyndall papers, photocopy of BN original. Note that Tyndall has not yet given up the tactic of attacking Bastian's skill (despite just announcing the heat-resistant spores), but has shifted toward expressing mostly anger at the fact that the younger man still outflanks him in rhetorical tactics.

86. Pasteur to Bastian, 11 Jan. 1877, included in letter to Tyndall, 12 Jan. 1877, Tyndall papers 27/C3.19.
87. Tyndall to Pasteur, 23 Feb. 1877, Tyndall papers, photocopy of BN original.
88. Pasteur to Tyndall, 24 Feb. 1877, Tyndall papers 27/C3.25
89. Pasteur, *Lancet* 1 (3 Mar. 1877); see also "The Spontaneous Generation Question," *Nature* 15: 380–381 (1 Mar. 1877).
90. Pasteur to Tyndall, 1 May 1877, Tyndall papers 27/C3.27; Tyndall to Pasteur, 5 May 1877, Tyndall papers 27/C3.27; photocopy of BN original. The bizarre confrontation that eventually took place between Bastian and Pasteur before the commission of the French Academy is described by Bastian, "The Commission of the French Academy and the Pasteur-Bastian Experiments," *Nature* 16: 276–279 (1877).
91. See Newton Nixon, *North London or University College Hospital: A History of the Hospital, from its Foundation to the Year 1881* (London: Henry King Lewis, 1882); also H. Hale Bellot, *University College London, 1826–1926* (London: University of London Press, 1929).

Conclusions

1. William Coleman, "Cell, Nucleus, and Inheritance: An Historical Study," *Proceedings of the American Philosophical Society* 109: 124–158 (1965), pp. 124, 127.
2. Alison E. Adam, "Spontaneous Generation in the 1870s: Victorian Scientific Naturalism and its Relation to Medicine," Ph.D. diss., Sheffield Hallam University, 1988, p. 150.
3. Ruth Barton, "Evolution: The Whitworth Gun in Huxley's War for the Liberation of Science from Theology," in D. Oldroyd and I. Langham, eds., *The Wider Domain of Evolutionary Thought* (Dordrecht: Reidel Pub., 1983), p. 279.
4. See Steven Shapin and Simon Schaffer, *Leviathan and the Air Pump* (Princeton: Princeton University Press, 1985); and Robert Kohler, *Lords of the Fly: Drosophila Genetics and the Experimental Life* (Chicago: University of Chicago Press, 1994).
5. Ruth Barton, "Just Before *Nature:* The Purposes of Science and the Purposes of Popularization in Some English Popular Science Journals of the 1860s," *Annals of Science* 55: 1–33 (1998).
6. Emil Duclaux, *Pasteur: Histoire d'un Ésprit* [1896], English trans. by Erwin Smith and Florence Hedges as *Pasteur: The History of a Mind* (Philadelphia: W. B. Saunders, 1920), p. 141.
7. Bruno Latour, *The Pasteurization of France*, English trans. by Alan Sheridan (Cambridge, Mass.: Harvard University Press, 1988).

8. George Wallich, "On the Rhizopoda as Embodying the Primordial Type of Animal Life," *MMJ* 1: 228–235 (1869), p. 235.
9. Hubert Lechevalier and Morris Solotorovsky, *Three Centuries of Microbiology* (New York: Dover, 1966), p. 36.
10. Jan Sapp, "What Counts as Evidence, or Who Was Franz Moewus and Why Was Everybody Saying Such Terrible Things About Him?" *History and Philosophy of the Life Sciences* 9: 277–308 (1987).
11. Martin Rudwick, *The Great Devonian Controversy* (Chicago: University of Chicago Press, 1985), p. 438.
12. George Wald, writing in 1954, actually attempted to resuscitate the term "spontaneous generation," precisely to revive the essential linkage with Darwinian evolution that Bastian always insisted on. Wald lamented the impoverished, ahistorical thinking of our times that made it possible for such a linkage to have been forgotten. See "The Origin of Life," *Scientific American* 192 (Aug. 1954): 44–53, p. 46. A December 1956 symposium Wald helped to organize on the origin of life was thus called "Modern Ideas on Spontaneous Generation." But opinion in the research community was so strongly against the implications of the term "spontaneous generation" that its usage essentially stopped at that point. Sidney Fox, the one prominent researcher to continue using the term, is the exception that proves the rule, especially since he had become significantly marginalized by the early 1980s.
13. See William Bulloch, *History of Bacteriology* (Oxford: Oxford University Press, 1938); also A. I. Oparin, *The Origin of Life,* English trans. by Sergius Morgulis (New York: Macmillan, 1938), pp. 20–26, 45–48. Numerous more recent authors have followed this highly influential account. That has led to much confusion, as it is historically naïve, even inaccurate; for instance in Oparin's insistence that most or all spontaneous generation supporters were vitalists (pp. 30–31). While he later points out that Bastian was an exception (p. 45), he completely overlooks Pasteur's fundamental vitalist tendencies and fails to note that both Bastian and Nägeli were extreme mechanists, even more so than Huxley. Generations of biologists have thus been left with the general impression that support for spontaneous generation is pretty much tantamount to belief in vitalism, though this is deeply paradoxical for anyone that accepts archebiosis or abiogenesis, rather than heterogenesis only. For a thoughtful analysis of Oparin's ideas, see Loren Graham, *Science, Philosophy and Human Behavior in the Soviet Union* (New York: Columbia University Press, 1987), chapter on "Origin of Life."
14. Another important historical work that reconstructs some of the continuity between Pasteur and Oparin is the Ph.D. dissertation of Harmke

Kamminga, *Studies in the History of Ideas on the Origin of Life, from 1860* (University of London, 1980). Much of this is condensed in her article "Historical Perspective: the Problem of the Origin of Life in the Context of Developments in Biology," *Origins of Life and Evolution of the Biosphere* 18: 1–11 (1988). John Farley discusses Bastian's later works in *The Spontaneous Generation Controversy from Descartes to Oparin* (Baltimore: Johns Hopkins University Press, 1977), pp. 152–154. From 1900 until his death, Bastian published four books on his experiments: *Studies in Heterogenesis* (London: Methuen & Co., 1904); *The Nature and Origin of Living Matter* (London: Fisher Unwin, 1905); *The Evolution of Life* (London: Methuen & Co., 1907); and *The Origin of Life: Being an Account of Experiments with Certain Superheated Saline Solutions in Hermetically Sealed Vessels* (New York: G. P. Putnam, 1911). The unavailability of Bastian's works has been a major obstacle to a serious re-evaluation. Fortunately, a number of them are being brought back into print shortly by Thoemmes Press, Bristol, U.K. in a set titled, "Evolution and the Spontaneous Generation Debate." The works reprinted will include *Modes of Origin of Lowest Organisms* (1871), *The Beginnings of Life* (1872), *Evolution and the Origin of Life* (1874), *The Nature and Origin of Living Matter* (1905), and *The Evolution of Life* (1907), as well as a selection of some of Bastian's early papers and the more important reviews of his books. The six-volume set is scheduled for April 2001 publication.

15. E. Ray Lankester, "Inaugural Address to the B.A.A.S.," *Nature* 74: 321–335 (2 Aug. 1906), esp. pp. 328–329; E. A. Schäfer, "Inaugural Address to the B.A.A.S.," *Nature* 90: 7–19 (5 Sept. 1912). Schäfer's address is also published as "The Nature, Origin and Maintenance of Life," *BMJ* 2: 589–599 (14 Sept. 1912). The *BMJ* also printed the discussion that occurred at the BAAS meeting after the address and, later, a rebuttal by Bastian: see "The Origin of Life," *BMJ* 2: 722–723 (21 Sept. 1912) and Bastian, "Further Experiments Concerning the Origin of Life," *BMJ* 2: 1542–1547 (30 Nov. 1912).
16. Steven J. Dick, for instance, looks at Schäfer's address in some detail in his book *The Biological Universe* (Cambridge: Cambridge University Press, 1996), pp. 337. So does John Keosian in his widely read *The Origin of Life* (New York: Reinhold, 1964), pp. 3–4, 114.
17. Bulloch, *History of Bacteriology*, pp. 108–109; Hubert Lechevalier and Morris Solotorovsky, *Three Centuries of Microbiology* (New York: McGraw-Hill, 1965), p. 36.
18. One such paper, sent by Bastian to the Royal Society in 1905, was read at the meeting of 16 March 1905 but then filed in the Royal Society Archives without even being refereed for publication. See "On the Heterogenetic

Origin of Certain Ciliated Infusoria from the Eggs of the Rotifer," 28 pp. ms., AP.76.7, RS Archives. Another paper, sent in 1902, was rejected outright. Bastian's comment was: "Were the announcements in this paper new or were they true? That they were new, there could be no doubt—their truth could only be gainsaid by investigation of the specimens." The Society also refused to accept Bastian's 1910 book *The Origin of Life.* He published the book himself in 1911, saying (p. 13): "What is the object of the Society but to advance Natural knowledge? And how can it expect to do this if it tries to stifle or ignore that which is adverse to generally accepted beliefs?" Apparently, since he continuously experienced stonewalling at the Society since 1873, Bastian must have felt himself an essential goad to keep the Society honest on such points of principle. Given the full context of his encounters with the Royal Society as I have outlined it, this interpretation makes more sense than seeing Bastian as merely irrationally obdurate.

19. See R. T. H., "Obituary of Dr. H. Charlton Bastian, F.R.S.," *Nature* 96: 347–348 (25 Nov. 1915); also F. W. Moth, "Obituary: Henry Charlton Bastian (1837–1915)," *PRSL* 89: xxi–xxiv (1917), esp. p. xxiii; and Edwin Clarke, "Henry Charlton Bastian," in C. C. Gillispie, ed., *Dictionary of Scientific Biography* (New York: Scribners, 1970).
20. For use of Huxley's "abiogenesis" after 1880, see Karl Pearson [1892], *The Grammar of Science* (London: J. M. Dent, 1937). Pearson was a good friend of E. Ray Lankester.
21. See also the article "Abiogenesis," by Peter Chalmers Mitchell in 1910 (*Encyclopedia Britannica,* 11th ed., v. 1: 64–65), in which Mitchell's attempt to redefine "archebiosis" in terms more conducive to Huxley's views indicates that Bastian's use of this term was still perceived to need a reply.
22. Schäfer, "Inaugural Address," p. 10. He also takes to task (p. 9) the Liverpool biochemistry professor Ben Moore, who took up some of Bastian's theory of colloids. Moore went on to publish a complete colloid theory of the origin of life in his *The Origin and Nature of Life* (London: Williams and Norgate, 1913).
23. On Loeb, see Philip Pauly, *Jacques Loeb and the Engineering Ideal in Biology* (London: Oxford University Press, 1987), esp. pp. 114–117. On Svedberg, see Anders Lundgren, "The Ideological Use of Instrumentation: The Svedberg, Atoms, and the Structure of Matter" in *Center on the Periphery* (Canton, Mass.: Science History Pub., 1993), pp. 327–346.
24. Marcel Florkin, "The Dark Age of Biocolloidology," in M. Florkin and E. Stoltz, *Comprehensive Biochemistry,* v. 30 (Amsterdam: Elsevier and Co., 1972).
25. Farley's chapter 8 takes a more historically sophisticated approach to the

"age of biocolloids." A more recent and still more developed analysis, particularly as relates to origin of life ideas, can be found in William Summers, *Felix D'Herelle and the Origins of Molecular Biology* (New Haven: Yale University Press, 1999), chapter 7. For more systematic attempts to recover the continuity between "biocolloidology" and nineteenth-century ideas on structure/function relationships in biochemistry, see Robert Olby, "The Significance of Macromolecules in the Historiography of Molecular Biology," *History and Philosophy of the Life Sciences* 1: 185–198 (1979), and Olby, "Structural and Dynamic Explanations in the World of Neglected Dimensions," in T. J. Horder, J. A. Witkowsky, and C. C. Wylie, eds., *A History of Embryology* (Cambridge: Cambridge University Press, 1986), pp. 275–308; see also Neil Morgan, "The Strategy of Biological Research Programmes: Reassessing the 'Dark Age' of Biochemistry," *Annals of Science* 47: 139–150 (1990). These papers bridge the gap between the "particle theories" of the 1860s and 1870s and theories of colloid chemistry of the origin of life in the 1920s and 1930s. Also crucial to this period are theories of viruses and their relation to the origin of life, e.g., in Thomas Rivers, "Spontaneous Generation and Filterable Viruses," *Northwest Medicine* 29: 555–561 (1930). This literature is very well reviewed in Scott Podolsky, "The Role of the Virus in Origin of Life Theorizing," *Journal of the History of Biology* 29: 79–126 (1996).

26. See Bastian, "Experimental Data in Evidence of the Present-Day Occurrence of Spontaneous Generation," *Nature* 92: 579–583 (22 Jan. 1914), and Bastian, "Use of Tyrosine in Promoting Organic Growth," *Nature* 95: 537–538 (15 July 1915). This was Bastian's last publication before his death on 17 November 1915. See also M. Wainwright, "Microbiology Beyond Belief?" *Society for General Microbiology Quarterly* 21: 95–99 (1994), and Wainwright, "The Neglected Microbiology of Silicon: from the Origin of Life to an Explanation for What Henry Charlton Bastian Saw," *Society for General Microbiology Quarterly* 24: 83–85 (1997).
27. On Herrera's support of Darwinism, analogous to Haeckel's role in Germany, see Roberto Moreno, "Mexico," in Thomas Glick, ed., *The Comparative Reception of Darwinism* (Chicago: University of Chicago Press, 1987), pp. 346–374, esp. pp. 346–350, 368–370. In his preface to this book (pp. xxv–xxvi), Glick also discusses the Italian figure who played a similar role as advocate of Darwinism and spontaneous generation, Paolo Mantegazza.
28. Alexander Oparin, *The Origin of Life on the Earth*, 3rd ed., English trans. by Ann Synge (Edinburgh: Oliver and Boyd, 1957).
29. See Sidney Fox, "Spontaneous Generation, the Origin of Life, and Self-Assembly," *Currents in Modern Biology* 2: 235–240 (1968); Fox, *The Emer-*

gence of Life (New York: Basic Books, 1988). See also Antonio Lazcano, "Origins of Life: The Historical Development of Recent Theories," in Lynn Margulis and Lorraine Olendzenski, eds., *Environmental Evolution: The Effects of the Origin and Evolution of Life on Planet Earth* (Cambridge, Mass.: MIT Press, 1992), pp. 57–69; Alicia Negrón-Mendoza, "Alfonso L. Herrera: a Mexican Pioneer in the Study of Chemical Evolution," *Journal of Biological Physics* 20: 11–15 (1994); and Stanley Miller, William Schopf, and Antonio Lazcano, "Oparin's *Origin of Life:* Sixty Years Later," *Journal of Molecular Evolution* 44: 351–353 (1997).

30. On Nägeli's ideas, see Pauline Mazumdar, *Species and Specificity* (Cambridge: Cambridge University Press, 1995); also Reinhard Beutner, *Physical Chemistry of Living Tissues and Life Processes* (Baltimore: Williams and Wilkins, 1933), pp. 119–127.
31. See Thomas Brock, *Milestones in Microbiology* (Englewood Cliffs, N.J.: Prentice Hall, 1961), pp. 100–101. This work available as a reprint from the American Society for Microbiology Press, 1999.
32. One of the first to make this observation explicitly was Theobald Smith. See "Koch's Views on the Stability of Species among Bacteria," *Annals of Medical History* 4: 524–530 (1932). Another was Polish bacteriologist Ludwik Fleck in 1934. See the English translation of his work, *The Genesis and Development of a Scientific Fact,* by Fred Bradley and T. J. Trenn (Chicago: University of Chicago Press, 1979), pp. 29–30. Fleck points out that the epistemological barriers to seeing the phenomena of asymptomatic carriers and bacterial variability were related and were both reinforced by the long shadow of Robert Koch.
33. Olga Amsterdamska has made a very illuminating study of the reasons why the "pleomorphist" position could be revived in the 1920s–1940s in "Stabilizing Instability: The Controversy over Cyclogenic Theories of Bacterial Variation during the Interwar Period," *Journal of the History of Biology* 24 (1991): 191–222. See also Harris Coulter, *Divided Legacy: Medicine and Science in the Bacteriological Era* (Berkeley, Calif.: North Atlantic, 1994), on the Kendall vs. Zinsser debate; and M. Wainwright, "Extreme Pleomorphism and the Bacterial Life Cycle: A Forgotten Controversy," *Perspectives in Biology and Medicine* 40: 407–414 (1997).
34. On the electron microscope debates, see Nicolas Rasmussen, *Picture Control: The Electron Microscope and Biology in America, 1940–1960* (Stanford, Calif.: Stanford University Press, 1997). Three papers by Rasmussen develop aspects of some debates in more detail: "Facts, Artifacts, and Mesosomes," *Studies in the History and Philosophy of Science* 24: 227–265 (1993); "Making a Machine Instrumental," *Studies in the History and Philosophy of Science* 27: 311–349 (1996); and "Mitochondrial Structure and

the Practice of Cell Biology in the 1950s," *Journal of the History of Biology* 28: 381–429 (1995). See also James Strick, "Swimming Against the Tide: Adrianus Piper and the Debate over Bacterial Flagella, 1946–1956," *Isis* 87: 274–305 (1996).

35. See Lynn Margulis and Dorion Sagan, *Slanted Truths* (New York: Springer, 1997), e.g., chapter 4, Margulis and Michael Dolan, "Swimming Against the Current," and chapter 20, "Big Trouble in Biology."
36. It is so constant that N. W. Pirie has cynically remarked that the eminence of a great scientist can be measured by how long he holds up progress in his field after his death. See "Concepts out of Context," *British Journal for the Philosophy of Science* 2: 269–280 (1952). For Pirie's observation that this applied to Bastian's work, see "Some Assumptions Underlying Discussion on the Origins of Life," *Annals of the New York Academy of Sciences* 69: 369–376 (1957), p. 370. Ludwik Fleck has much to say about how this process operates. See his *Genesis and Development of a Scientific Fact*, pp. 26–27, for example.
37. More stories in this vein, from c. 1930–1970, are briefly reviewed in Adolph Smith and Dean Kenyon, "Is Life Originating *De Novo?*" *Perspectives in Biology and Medicine* 15: 529–542 (1972). For others, see Carl Lindegren, *The Cold War in Biology* (Ann Arbor: Planarian Press, 1966), especially chapter 8 on the origin of life.

SOURCES

Archival Materials

American Philosophical Society (APS) Library, Philadelphia: Darwin papers, Owen papers, Bastian letters, and Wyman letters.

British Library: Alexander Macmillan papers, Richard Owen papers, Lord Avebury (John Lubbock) papers, miscellaneous assorted correspondence (Robert Brown and others).

British Museum of Natural History (BMNH) Archives: Botany—Robert Brown papers, Zoology—Richard Owen papers, Ray Society papers, T. R. Stebbing letters.

Columbia University Library, New York: Frederick Barnard papers.

Correspondence with surviving Bastian descendants, Miss Julia Bastian and the Dowager Lady Ann Rathcreedan.

Countway Library of Medicine, Boston, Mass.: Jeffries Wyman papers.

Exeter University Library, Exeter, Devonshire: Norman Lockyer papers.

Imperial College, London: T. H. Huxley papers.

Johns Hopkins University Archives: Tyndall and Huxley letters.

Kroch Library, Cornell University, Ithaca, N.Y.: Burt Wilder papers.

Linnean Society Library, London: James Murie papers, H. C. Bastian letters.

National Hospital Queen Square Archives: Bastian papers and casebooks, and Institute of Neurology Library, London.

Reading University Library: Alexander Macmillan papers.

Royal Botanical Gardens (RBG) Archives, Kew: Joseph Dalton Hooker papers, William Thistleton-Dyer papers, George Bentham papers.

Royal College of Physicians (RCP) Archives, London: Charles Murchison papers, miscellaneous correspondence in autograph letters collection.

Royal Institution, London: John Tyndall papers. Also includes T. A. Hirst papers, X Club notebooks, photocopies of Tyndall to Pasteur letters from Bibliothèque Nationale, Paris.

Royal Society (RS) Archives, London.

University College London (UCL) Archives: John Burdon Sanderson papers, E. Ray Lankester papers, Henry Charlton Bastian papers, Francis Galton papers.

University of Edinburgh Library: John Hughes Bennett papers.
University of London Library, Senate House, London: Herbert Spencer papers.
Wellcome Institute for History of Medicine, London: Edward Sharpey-Schafer papers (including Physiological Society notebooks), Bastian letters, assorted miscellaneous letters in Pasteur papers, autograph letters collection.

Unpublished Materials

Adam, Alison E. "Spontaneous Generation in the 1870s: Victorian Scientific Naturalism and Its Relation to Medicine." Ph.D. diss., Sheffield Hallam University, U.K., January 1988.

Barton, Ruth. "The Purposes of Science and the Purposes of Popularization: Some Popular Science Journals in the 1860s," presented at Royal Institution, London, 21 September 1994.

Bynum, William. "Sir Joseph Fayrer: The Natural History and Epidemiology of Cholera," presented at University College London, 3 October 1994.

———. "The Evolution of Germs and the Evolution of Disease: Some British Debates, 1870–1900." Dibner Institute conference, "Pasteur, Germs and the Bacteriological Laboratory," Cambridge, Mass., November 1996.

Gruber, Jacob. "The Richard Owen Correspondence: An Annotated Calendar." Unpublished ms., American Philosophical Society Library, Philadelphia.

Haas, J. William. "Late Victorian Gentlemen of Science: William Dallinger and the Wesley Scientific Society," presented at Royal Institution, London, 19 September 1994.

Kamminga, Harmke. "Studies in the History of Ideas on the Origin of Life from 1860." Ph.D. diss., University of London, November 1980. Dr. Kamminga kindly shared the chapters relevant to the debates of 1860–1880 with me.

Mendelsohn, J. Andrew. "Cultures of Bacteriology: Formation and Transformation of a Science, 1870–1914." Ph.D diss., Princeton University, January 1996.

Rang, Mercer. "The Life and Work of Henry Charlton Bastian." University College Medical School, London, 1954. A copy of this master's thesis was kindly given me by Dr. Rang.

Romano, Teresa. "Making Medicine Scientific: John Burdon Sanderson and the Culture of Victorian Science." Ph.D. diss., Yale University, May 1993.

Uyterhoeven, Sonia. "Protoplasm and the Reform of Physiology." Master's thesis, Wellcome Unit for History of Medicine and Department of History and Philosophy of Science, Cambridge, U.K., December 1991.

Worboys, Mick. "'Deeper than the Surface of the Wound': Lister, Surgeons and Germs in Britain, 1867–1885." Dibner Institute conference, "Pasteur, Germs and the Bacteriological Laboratory," Cambridge, Mass., November 1996.

INDEX